Multiplying Your Genius Within

Sergio Antonio Meneghetti

Pindamonhangaba – São Paulo – Brazil

June 2025

Second Edition

Cover: Sergio Antonio Meneghetti

Editing: Sergio Antonio Meneghetti

Translation: Marcio Bonetti

To the reader:

In this second edition, it was necessary to add new information and important cases, mainly in the scientific area. This book is my best-selling work in the United States.

Intuition is a fact. It is present in all the minds of humanity, to a greater or lesser degree.

Civilization is in the phase of consciousness (Reason and Analysis) and moving towards superconsciousness (Intuition and Synthesis = Genius).

These lines show intuition in scientific and practical forms and the path to developing your inner Genius.

There are more than thirty years of research and experiences in the intuitive area.

Intuition acts in all areas of human knowledge, it is a universal phenomenon that anticipates the future. The future comes from the human mind.

This is the most advanced process of perception and research that human beings have, because it occurs in the spirit. This allows human beings to feel and visualize phenomena beyond material perceptions, known as the five senses.

The objective of this book is to show the way to understand how the intuitive phenomenon works, that it is something logical and scientific and that Genius is a personal achievement through effort and improvement.

Sumário

Introduction:

I am a person of intuition since my youth, I researched and I am also object of research on the subject.

This work is aimed at all professional, businessmen, entrepreneurs, students and researchers.

This work is intended for all professionals, businesspeople, entrepreneurs, students, researchers and, of book, people seeking knowledge.

What I share with you is the result of personal experience in work and life.   In the book of this work we will see cases demonstrating what is proposed for our professional growth and in the researches. It is a totally different way of absorbing knowledge.

Cases such as:

Approval of major project in the self- motivation area.

Decision making vital to an important business.

Attention to intuition that saved my life of an imminent accident.

Receive an email reply (35 lines) from President Barack Obama.

Ideas that have led to improvements savings in my professional environment.

- Important views and perceptions in the scientific field.

For what I will submit, you will come across to completely different concepts to the traditional, and this is for a simple reason: the human race is reaching a spiritual and intellectual maturity necessary to understand this phenomenon scientifically, which not have been possible in the past. It is by this path that I want to help you to improve and progress in all areas.

If you want to be an entrepreneur, you will see the best path. If you are a professional, owner of a company or business, here you will see that intuition is an important tool in the achievement of your goals. If you are a student, you are already getting ready to be included in the market with something more to offer.

If you are a researcher, you can be sure; this is the most advanced tool that exists in the area of research.

Notice something important: "Everything that was, is and will be built by human is born in the mind".

The expansion of your mind will be propelled for your growth in all areas.

In this paper I will use minimal figures or photos so you can put your mind to evolve, after all thought in action is internal growth. Another reason is sustainability, because you will have a cleaner and economical book, if you want to make an impression.

I will speak of complex things about the universe and life, but don´t you worry, this information is needed only to show the steps of evolution, until you reach the focus of our study.

Another important factor is: Here you will come across completely new concepts, and this may cause some difficulty understanding, but I can´t show a new way for old roads.

I will help your mind to open up in order to uncover new possibilities, and so you will be prepared to "create and innovate" in your life and in you work.

Start to imagine that everything you are used to seeing "can be done differently". Any work or life style.

I must warn you that this book deals with very advanced concepts, and it is a clear and growing synthesis on this "new" subject intuition or insight, also known as sixth-sense.

I can assure you that it is a different and exciting road.

At first it seems to be only the fragment of an idea, but when one pays attention and seeks to develop the idea, it grows so surprisingly.

You will begin to understand a unique format to acquire knowledge and most important:

"This knowledge will come from within you.

It is worth remembering that most of the new and revolutionary knowledge happened in a non- traditional way, like real lighting in the brilliant and trusting minds.

"It is not geniuses that reveal the genius within, but the genius within that reveal the geniuses".

In other words, it is about people who reveal themselves every day.

You can be a great source of these ideas; it is simple as enough desire to put on an effort no matter the size of the idea.

Riches and victories will be a result of the effort of each individual.

You can be sure of one thing: You will face the world in a more complex, perfect logic way.

Being aware of this fact, you already took a huge step in relation to the average of civilization.

Today I can transmit information that I have already experienced through this psychic channel, including previously unpublished knowledge. The details that I present are the result of visions and studies and are included in other specific works.

Among the works cited, I must present the basic book that was fundamental for my understanding:

"The Great Synthesis" by Pietro Ubaldi, written through an intuitive process between 1931 and 1935.

Presentation

Perceptual Scales

Genius

Where

Why

Business

Cases

Personal Preparation Tips

Acknowledgement

Sergio Antonio Meneghetti

Perceptual Scales

The subject of perception that will be addressed in this book is the key to the progress that awaits us.

Over time, humanity has forgotten some of its potential by imagining or placing all its possibilities in the materiality of things. In this way, man has lost the address of his inner potential, who he really is, and how he could advance in this sense.

The traditional learning process has generated many advances to date, but for new challenges, as well as the dependence on technologies, phenomena, whether physical, chemical, biological, or others, will require new tools for their understanding and observation.

These technological revolutions to drive progress happen, as is the case with the computer, which, even when only data storage is in mind, has already avoided files with millions of pages. In other words, for major challenges, new, more efficient, and compact ways to solve the situation emerge.

In the case of life, this evolutionary process also occurs, and this starts from the rustic to the more refined and complex systems.

Since this book does not intend to be in-depth on certain subjects, as the amount and sources of information would be large and would distort the intention of the work, I will present some information in a concise way.

The graph below shows the evolution of perception in a simple way.

Evolution of Perception

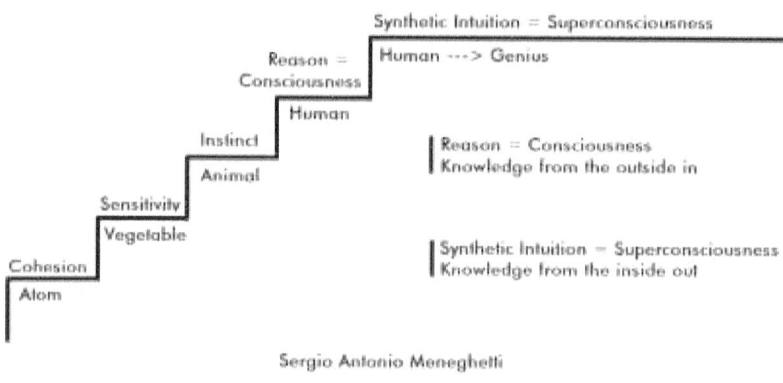

Figure 01

Below each step of the ladder is manifestation, and above it is the process of perception.

- Contact between atomic individualization will perceive another atom through the cohesive phenomenon.

- In the next evolutionary phase, in this case, the plant phase, the system is already more refined and evolved, and in this initial phase of life, the contact of information with other forms occurs through sensitivity.

- When life climbs another step and reaches the animal phase, the process of perception with the external acquires the characteristic of instinct.

- Continuing the evolutionary climb, the animal phase reaches the point that would revolutionize the entire existential environment, and in this phase, progress occurs in a much more complex and organized way, because at this moment, the human being acquires consciousness.

- Currently – and I am not referring only to the years in which we live, but to the last millennia – civilization is in the transitional phase between consciousness and superconsciousness.

In the four previous phases, the means of perception used the material environment as an instrument for transmitting and capturing information – the best known of these being touch, hearing, taste, smell, and sight.

In the case of superconsciousness or Intuition, the process changes in quality, moving from material interaction to a more abstract interaction. In this evolutionary phase, the spirit becomes more present as an instrument of perception.

It is paramount to remember that this is a simplification and that, when we talk about spirituality, we are addressing it in a more scientific rather than religious way.

With the refinement of perception and its complexity, it would be important for human beings to know more about spirituality and its mechanisms. This is one reason why I have addressed it, even if not in depth.

In short, the pineal gland acts as an antenna, an instrument that receives information that manifests itself in other vibratory fields of the universe.

At this stage, this very delicate instrument captures the information, generally through electromagnetic waves, and the brain does the work of processing the information.

From the outset, the reader comes across concepts that have not yet been covered by contemporary science, given that doubts still arise regarding the evolutionary scale. Even with the work of Charles Darwin, there are still questions to be answered, and, little by little, science is building this guiding thread.

Despite being an evolutionary ladder with apparent gaps or chasms between one step and another, we must bear in mind that, according to scientific logic, from the moment the connection between one phase and another is broken, evolution also breaks. It would be like a long ladder cut into pieces, thus forming smaller ladders.

There must, therefore, be a phenomenal mechanism that promotes continuity between the phases, even if we do not yet have knowledge of these processes.

Another important and controversial piece of information about this evolutionary continuity is that, in the case of life, for example, the death of the body does not destroy the essence that controls it, that is, the spirit or the self.

According to the laws of physics, nothing generates nothing, so if something exists, it will always exist in some form.

With these few clarifications, we can deduce that, if evolutionary rupture cannot occur, the human being is a repository of information generated throughout all his existences in this growing and complex ladder.

Philosophically speaking, our intellectual, spiritual, and biological potential is the fruit of ourselves. This can explain the differences in individual potential.

To my dear reader, it is already a little clearer that you have latent potential within you, you just need to develop this dormant treasure.

Follows text:

Text of the work: "The Great Synthesis" by Pietro Ubaldi

You already see how the edifice that reason is capable of constructing can anticipate direct observation; this is just the vulgar path of a thought that always rests on facts. Imagine what discoveries you can quickly arrive at, when scientific problems are faced intuitively, as I told you. Incidentally, the true and great discoveries were all flashes of genius intuition, the superman of the future who, leaping beyond rational forms of research, anticipates the intuitive forms of future humanity. The great leaps forward were taken by man, never experimentally, never rationally, but by intuition, true and great research system of the future.

Intuition in the View of Science and Philosophy

- Albert Einstein – "There is no logical way to discover the laws of the universe, the only way is Intuition"

- Steve Jobs – "Listen to your Intuition"

- Nikola Tesla – He saw the Alternating Current Motor through an Intuitive process.

- Carl Jung – "Intuition is the heart in practice"

- Henri Poincaré – "We prove through logic, but we discover through intuition "

- Arthur Schopenhauer- "Intuition is not opinion; it is the thing itself"

- Joseph Joubert, - "Reason can warn us more than avoid it, only Intuition tells us what to do"

- James Balmes – "One of the characteristics of genius is Intuition: seeing without effort what others would only discover with great work"

- Immanuel Kant – All human knowledge began with intuitions, passed from there to concepts and ended with ideas.

- Immanuel Kant – Thoughts without content are empty; intuitions without concepts are blind.

- Epicurus – It is true both what we see with our eyes and what we learn through mental intuition.

2

Dimensions

Having understood a little about the process of evolution and creation of life, this evolutionary line will now be placed on a dimensional ruler, or on an increasing scale, relating the phenomena to the different dimensions.

The vast majority know that the first spatial dimension is the expression of a line, or casually speaking, a straight line connecting one point to another.

The second spatial dimension has the feature of volume, that is, the surface connecting the points perpendicular to this (side x side x height).

For those who are familiar with scientific language, can have a better notion imagining a line on the floor of your house (first dimension) or the floor of your house (second dimension) and a room in the house that contains the floor, wall and ceiling (third dimension).

These dimensions may vary in size, and in some way can be measured.

Now, complicating a bit, the next dimensions cannot be

spatial as the volume would involve any other imaginable type of spatial dimension.

The next dimension, or the fourth dimension, is already known by science as time, and in this case, the dimension is no longer something in space nor does it have a feature, or something that is more noticeable in the mind than in anything concrete.

According to The Great Synthesis, this sequence is a new dimensional trinity, or three dimensions with conceptual characteristics.

Therefore, time was born with the degradation of matter in its first formation and this degradation generated a transmission speed of the phenomenon.

This transmission speed has a Linear feature (first conceptual dimension).

Just as in the second spatial dimension the surface is constructed by lines, in the second conceptual dimension the process is similar, repeating the same model.

The evolution of this psyche after several phases came to state consciousness (reason and analysis), the current phase of the Earth´s population. At this stage, reason and analysis has the characteristic of surface, or it reasons a situation, does an analysis and concludes by generating a result. In this case, it can be said that this is manifested in the second conceptual dimension.

The next big step of the human being is in the third dimension, that is, the super-consciousness (intuition and synthesis-genius). Vision and understanding of a situation or phenomenon globally and instantaneous. The situation has volume features.

To make it easier to understand, let´s make a comparison.

Imagine you are sitting watching the presentation of a symphony. You will see the conductor, the musicians and all objects that are in the place of fractional form, that is, your brain observing, thinking and analyzing all the information in this way, you will have after some time the idea of the set, including sound.

Now, imagine you high above the orchestra, with his eyes closed. From the moment you open your eyes, you will have an instant view of this orchestra so, at a minimum time you will have all he information about the subject, almost out of time and space.

In this second case, the brain realized the whole set in a synthetic form and instantly.

This is a rather rough example, but is more or less how you process the information in this third conceptual dimension.

Another example, which many may have had within their homes, is the following:

Your mother or another family member knows or feels that something is going on with another member of your family, which is distant. Usually from mother to child. Unfortunately some unpleasant things.

In this situation, the mother realizes something is happening, without seeing or observing the phenomenon. This event is registered outside of time and space, if we can define it so.

In this way, you can assemble and score on evolutionary rule, where the intuition and the synthesis are in the evolution of the cosmos and of human beings.

Spatial Dimensions:

First Spatial Dimensions (Line)

Second Spatial Dimension (Surface)

Third Spatial Dimension (Volume)

Conceptual Dimensions:

First Conceptual Dimension (Time-Line)

Second Conceptual Dimension (Consciousness-Surface-current phase)

Third Conceptual Dimension (Super-consciousness-Volume-genius)

This is the great scientific base that the scholars of intuition are unaware of because in their vast majority, just study the phenomenon externally and indirectly, in other words, it is normal to hear about "insight" or intuitive manifestation. Knowing that something happens and manifests itself, however unaware of how and why of this process.

As intuition is observed or perceived by the spiritual instrument, it becomes more difficult to test the phenomenon directly and materially; therefore, the greatest proof of the existence of the intuitive phenomenon will be the facts resulting from this process, that is, there is the perception or vision of something, often outside of time and space, and, afterwards, the fact happens, endorsing the phenomenon as true.

3

Genius

The genius may be synonymous with personal talent, and every human being has a personal talent, to a lesser or greater degree.

This talent manifests itself in all areas, from the art to the simplest skills. Yes, individual manifestation is externalized in your enhanced form.

Genius is not a gift, but the result of the effort of every human being in your eternal walk.

In a nutshell, is the dilation of consciousness heading for a super consciousness.

Genius is a treasure to be developed by individual effort.

The sooner it becomes aware of this internal treasure, the faster the progress of the planet will be.

There is a lot of confusion between genius, intelligence and accumulation of information.

Genius is a creator, perceptive, insightful, and above all

altruistic.

Intelligence is the capability of reason, analyze, reflect and work situations.

The accumulation of information is the capability of archive, extract, inform and demonstrate content.

On Genius, the human being can be devoid of culture or academic knowledge, but can be genius within his intellectual size.

On intelligence, human beings work with ease the information at hand, take full advantage of this information by generating a new set leaving these, however, have difficulties creating new, outside of the acquired concepts.

In the case of human being well informed, for everything that you are asked will have the answer. But have difficulties generating or working this information, and create or accept something new.

There are no demerits under these conditions exposed, because every human being has this qualities and needs.

One should not label or belittle the different abilities or disabilities. Everyone has his noble role in the gear of progress.

By clarifying these differences, we can show that genius is intrinsically linked to the evolution of the human being.

No one will become a genius by buying brain pills at the pharmacy or by developing learning and memorization techniques, but this ability can and should be developed when the person already perceives some situations in which they deviate from normality.

The important thing in this context is for each person to

know themselves internally, to observe themselves, and to pay attention to their strongest thoughts, as if they were a warning or an idea of something new.

Observing the Perceptive Ladder, there is another interesting point: as in ordinary learning there is an accumulation of information, in the existential phases, from the atom to the genius, the information is also recorded throughout this process, that is, all the experience is registered in the being, or to be more direct, in the spirit, which is the eternal phase. In this way, when a vision or idea arrives – naturally, in those who experience this possibility – , they already have the intellectual material to be able to work with the new information.

Human beings do not need to be like the famous geniuses who change the course of history, but rather to achieve what they are capable of.

It is also important to note that the well-known geniuses were people with moral qualities and simple conduct. The social fruit was always the scope of their work.

Sergio Antonio Meneghetti

4

Where

Companies

Research

Environment

Arts

Armed Forces

Personalities

Daily Routine

Sergio Antonio Meneghetti

Companies

Companies are great fields for the use of intuition.

Personal, professional, diversity of processes, technologies and other factors, generate an environment of needs, and this tool can give the most varied responses, developing the potential of each member of the group.

The call for constant innovation and productivity, combined with the profit (life of the company),

Are springs that drive progress as a whole.

Many companies are born just about great ideas, as well as new products and processes.   The how to is essential to give greater momentum to the results.

The intuition or the development of individual genius, is not restricted to a professional, a team or area, but it acts at each point where needed, therefore, it is necessary in all the hierarchical line.   Everyone can contribute to the size of your performance.

Investing in human potential, in order to provide you

with a new vision of your inner potential, comes out a lot cheaper than the change of equipment, for example.

A good idea could avoid waste, save raw material, time, natural resources, occupation of areas and generate better products in all aspects.

In term of "security", which is of extreme importance, attention to intuition can avoid accidents, diseases and save lives.

In the chapter "cases" ahead, there will be valuable practical information demonstrating theses aspects.

Research

All this is essential for the smooth running of the scientific progress in all areas, but in a quick note, it can be observed that all this only becomes necessary after the birth or generation of great ideas.

Observe how much progress was generated the theory of relativity of Albert Einstein or Isaac Newton and other innumerable, in several areas of science.

In this chapter I would like to make a vital observation for the book, giving a different view of the matter "the importance of the fact". Speaking of geniuses, virtually all science pay attention to two factors:

First, the idea itself, and in its revolutionary content.

Second, the author of the idea, or the genius who knew how to work this valuable information.

What little science researches is the "how" this psychic process work. The "how" was born from this view or process of perception.

This is the scope of this book it shows how this process happen and what is needed to develop it.

This is a challenge for the researches:

To invest in technology and academic growth, or to invest in this line of abstract research? I would say that on the whole, investing in both will always be more productive.

Here is another note to science:

Science works based on reason and analysis, giving little room for intuition and synthesis.

This is not an observation in a derogatory sense, but intuition and synthesis are steps in the way of the future, this super-conscience is still undervalued by many who are unaware of their phenomenon's .

As science has dominion over the relative or "time and space" when it begins to manifest itself in the "Absolute", or something abstract without time and space, the rejection is natural, it get out of the consistency proof usual field.

Using the example of Einstein, testing the theory of relativity was difficult, but possible. However testing the psychic phenomenon that occurred in a fraction of seconds in your mind to realize the observation of the theory, is hardly possible.

Academic Environment

School is the environment that shapes the human race.

The plurality of knowledge walks in their hallways, fertile environment to feed the soul, inexhaustible barn.

No matter what corner of the world, the school will always be vital in the life of a human being.

Every location that brings human growth should be a blessed place as it is helping to build the greatest treasure that a human being can acquire: knowledge, and with this his unconditional freedom.

Nothing is as important as a mind free to fly in the field of abstract, generate thoughts and distribute this gift that comes from the universe.

The universe has all the possibilities and it is in the search of this that we must walk.

There is a secret science to be unveiled and assimilated in everything, and the veil is just our ignoring this treasure.

Every atom organized in the human body is a result of the millennial learning of the psyche. There is no disorder in the universe, all moves following an ascending path. Evolution does not stop at the relative.

Note the knowledge expressed in the tender ages transmitted with simple words, comparing the rustic way that a child can absorb. With the passage of time and the accumulation of luggage, the human being or the eternal student of life, will acquire conditions for more complex understanding.

As mind matures knowledge can be expressed in a profound and scientific form, i.e. the science demonstration of how the phenomenon happens

Today this is allowed because the human race has come a long way, so out of childhood as a civilization is ready longer and higher flights.

Man penetrated the intimacy of matter and already handles his energies, got his feet off the ground reaching the cosmos but still need to have the same bold detachment to launch himself in the moral field.

Without this balance, mankind can suffer effects caused by the lack of law that regulates everything.

By analogy, comparing the beginning of knowledge disseminate on the planet, it can be observed that that human race in the past, did not have enough understanding to be able to understand science or the mechanism of things as today. Somehow knowledge descended to the minds and with simple words and easy to understand the great truths were taught.

As the so-called prophets, saints, masters, oracles, elders, priests or others could know this knowledge that led the people of that time?

How were events or phenomenon explained in those days? Today much of that knowledge is brought to light by the scientific evidence.

What is the relationship between a prophecy and its confirmation sometime later?

Using the information in the chapter "Universe and Life" and "Dimensions" one can understand that only an evolved mind could be capable of intuition and synthesis. In other words, the intuition and the syntheses are abstract phenomenon out of time and space, so that phenomenon of vision of the future or anticipated knowledge can be explained.

If creation used these devices for the progress of this planet, so today we can have an idea of what awaits us with the development of this ability.

Is there anything more important in this context, perhaps most student don't understand, learning begins to char its source, i.e. Today´s learning comes from outside the human being, but with the expansion of consciousness or super-conscience the knowledge will come from the inside out.

Being more scientific, human beings not only will notice the phenomenon of the universe, but will begin to understand them participating in theses phenomenon.

I would be like quitting observing the dynamic movement of the water of a river externally and entering the river to feel and understand its dynamic more deeply.

Can you understand now some of the importance of this breakthrough in the sense of intuition and synthesis? As

Sergio Antonio Meneghetti

this breakthrough personal development can bring progress to the planet.

This matter was brought to this chapter for one simple reason:

How much these subjects progress would result if it was brought up since childhood?

It would create a strategic culture to the country.

How much progress for the nation that pays attention to this fact.

How technology would be generated on a sustainable basis.

How much natural resources would be spared and equilibrium for the environment.

This is not just a book to develop more experienced professional, but a wise development strategy for the planet.

In short: the academic environment can contribute even more to the wellbeing of its civilization.

Arts

Amongst all manifestation, abstract art is the most profound; it is the capture of universal raw material that nobody sees, however, most feels.

According the work The Great Synthesis, the pure artistic manifestation, is the one that comes closest to its Creator to mention the pure communion of prayer, which is a direct communication.

It is the perception that is born within the soul and materializes in different ways, from song to poetry, from painting to sculpture, from writing to acting, etc. It is the expression of beauty that unfolds every day in different ways.

Without the artists themselves being aware of this fact, it is the intuition at work, showing the intimacy of the inner being, showing the real treasures stored in each individual.

Down to materialism, how much money artistic manifestations generate.

Excluding the so-called pseudo art which are imperfect manifestations and according to the audience, accepted as art.

The real art remain in time. This is the best gauge of what is real and what is ephemeral.

True art elevates, touches feelings, encourage, cheers up, enlightens, carries the human being to feelings and sensations of bliss.

Let´s make a comparison:

If a human being using this same tool of the artist, in the various fields of science and life, in the workplace or in research, how much positive results would be achieved?

This information confirms the existence of conceptual dimension. Here science cannot use mathematics to be able to calculate this phase of the universe.

This is one of the reasons why official science has no answers to many questions about the mechanics of the universe. There is a fine line, however rigid which give continuity to the physical universe in the conceptual universe.

Another important observation that can demonstrate the formation of the physical universe is that the same came from something also conceptual. This reinforces the presence of something much bigger, the existence of a Creator (God).

Art is the main example of the manifestation of the power of intuition or genius.

Pure art is simple and conveys much information and sensations, overdone and excessive art overloads the senses and convey little information. Can you now understand the meaning of synthesis?

Synthesis is the practice of the future, that is, the wise economy o words, processes, methods, arts, expression, etc.

An example of a sentence which expresses very synthetic, and practically the good living comes down to this:

"Love God above all and your neighbor as yourself"

Apart from the religious sense of this phrase, let´s observe it on the practicality of everyday life.

In the professional environment the major consultants, speakers or scholars teach us to treat well our subordinates, customers and co-workers.

In international relations isn't it the same?

In politics, ethics and justice aren´t this connect to this maximum?

Finally in human relations isn´t this practice that generates the best results?

See how logical is the importance of this new conquest of mankind and how many benefits it has. The conquest of intuition and of synthesis.

Sergio Antonio Meneghetti

Armed forces

Perhaps few people know how many advances in science and research were created from wars, or more specifically in the armed force of most developed countries.

While civil society goes on unaware of much information, other institutions are attentive to everything that is new and can generate strategic knowledge.

Did you know that the army and navy of the United States spend millions in research on the use of intuition for military purposes?

Spies are prepared for intuitive situations to make the most of these?

In the case of the army and navy what I read on the subject would be about prevention in advance on the battlefield, in other words, the intuition would be used in the form of premonition of danger.

While the great mass of the population think that intuition is a girl thing or that thing is religion or exoteric philosophies, scientific institutions are going deep into the subject.

I want to make an important observation for those who want to use these achievements

"intuition and synthesis" for less noble purposes: Insight and synthesis a the achievements of human evolution and the primary factor is the moral evolution, not just scientific developments.

How intuition is the contact with the invisible plan and with the essence of phenomenon if it is uses with the intention of harm or for the purpose of power, of book the source will close and who tries to use it for these purpose will be fool by less evolved force falling to the detriment of their own. It is the price to be paid for those who want to avail themselves of something sublime for deplorable purposes.

If you are trying to save lives, this power will manifest itself positively.

If it is to destroy lives, it shall not manifest itself.

Personalities

If we were to mention in this book the personalities who used intuition (even without knowing) to bring progress to the planet, it would fill many pages.

Cited earlier, Einstein, Newton and so, many could be cited as geniuses of all time. In music, painting, sciences, literature, poetry, sculpture, singing and so on.

In this chapter it goes on record how this subject is important in people´s lives, and how this conquest brought progress and advancement for the planet Earth.

Thanks to contemporary genius you are having access to this material and becoming aware of a wealth of information from all corners of the world.

It is worth pointing out that all this is possible due to a seemingly bad factor, "necessity". Without it, little would be done for pure progress.

The greatest works generally were created from the pains of their creators. It is the price of progress, but it is worth it.

Nobody is exempt from being one of these personalities of progress, so you and anyone who apparently has no notoriety or titles can accomplish great and revolutionary works.

Believe in yourself, you are the one who knows more of your intimacy and potential. It is not easy but you can get there.

It will be an honor for me to see a student or reader to realize a work the helps in the improvement of this civilization.

The genius walks in anonymity.

What is your talent?

Daily Life

A castle is built only with large stone?

Of book not. So are the intuition and the synthesis, they manifest themselves in all areas of life and the universe.

All construction is composed of very small parts, and these manifest themselves to a lesser or greater degree, but all have its importance in the overall context.

The message is that genius can be in the simple thing each day in a small idea that causes the housewife to get less tired, or a tool that facilitates a gigantic construction, in a better way to perform an activity, a thought that prevents something unpleasant, and so on. There is no formula for this purpose.

Throughout all times of existence of this planet, the action of the psyche was based on all forms of constructions from protoplasm to the most sophisticated device, the wall of the insect to the conquest of space, everything was generated in the mind first.

The mind is your greatest treasure, in other words, you

are your greatest treasure because you are the mind, and the body is only a garment where you manifest yourself in this physical universe.

The mind works in the physical universe, creating the conceptual universe.

In your everyday life a great work is built at every moment, at every action taken.

You are very important to the world.

"In everyday life the greatest work of all time is realized, Life".

5

How

Intuition

Comparative

Moral

Nutrition

Environment

Sergio Antonio Meneghetti

Intuition

To speak of intuition as a genius tool, first of all it is important to have an idea of the importance of it, or a vision of the benefits that is has against reason and analysis.

Vision of Intuition:

While man looks at dead stones, trying to unravel old written life´s mysteries the new science bring crystal clear messages clearly and quickly through intuition.

The first is analytical and requires a lot of knowledge, economic, personal, technological effort and time.

The second is synthetic and requires the polishing of human morality, knowledge and personal effort.

The first requires analyses, interpretation and theories.

The second requires perception, interpretation and learning.

The first studies the past.]

The second launches man into the future.

The first is restricted in space and time.

The second is open to all.

The first is limited.

The second is unlimited.

The first is the ancient values of all residues.

The second is the essential juice to give life to the millennia to come.

The first is a step overtaken.

The second is the platform which rises.

The first made man.

The second will make you an angel.

The first took you from the clay in which you were created.

The second will show you the mysteries of heaven where you came from.

The first chained you to the principles.

The second will set you free by excelsior purpose.

Science, human labor, product ladder that leads to the knowledge of the intimacy of matter, it is the lens that allows you scour the universe micro and macro, showing us a whole and complex mechanism.

And when this science cannot give more answers, science from evolution of life will answer by internal processes where it will not be necessary to gather date to explain the phenomenon, but this process will be understood as a whole in a glimpse through the intuitive perception.

"Intuition is the science addressed to science because in it there is science".

This new level in which the human being begins with your foot on the evolutionary road, will bring new life on basis.

One of the first observations to that intuition can manifest itself, is to put aside reason and analysis. These comprehend the brain thinking and analyzing the facts and phenomenon, preventing the penetration of knowledge or perception.

Think about the subject with interest and let the mind free, it will go thru unusual paths ad find the answers.

Not always worth the effort because there are no fixed rules for the opening of this channel, so in unexpected moments the intuitive phenomenon can happen without the will of the human being.

Observe that in the history of great geniuses that the idea comes by chance.

The reason and analysis are blockers of intuition.

This is one of the reasons why science snags in certain concepts. Generally scientist by studying and by having this information luggage try to formulate ad deduct pure reasoning based phenomenon based purely on reasoning, but generally the new is outside these parameter of knowledge (why it is the new).

Sergio Antonio Meneghetti

Comparative

For a better understanding (and this is how humanity learns) we will make a comparison of the human mind with an electronic equipment.

The brain is a kind of a radio communicator, i.e. receives information and also issues information.

As evolution is law in the relative field, as everything evolves suffering progressive transformations, the radio in question was also born the rustic and noisy in order to turn into another evolutionary type of transmitter and receiver. With this quick delta science in recent times, you can verify the real revolution in these electronic instruments.

Yesterday it was a wooden box with large parts and high energy consumption. So the transmission and the reception were imperfect. Today the equipment is quite smaller and the sound refined and perfect.

For this to happen, it was necessary human effort, new parts, better conductors and of higher purity, and so making it possible to decrease the size of the equipment.

Those were some of the changes, but there were other more important and should be observed, that is, the concept of the new equipment. There was the development of the equipment until a saturation point of that technological model and only a revolution of that type of technology can give new life and longevity in this type of transmitter and receiver, thus creating a more improved performance.

The variation in evolution of life has a longer cycle, but it happens.

Compare a primate, with a body of the present day, it allows us to have an idea of change or human organic machine intended.

Since it is not possible to evaluate a brain by remote vital dates (birth, growth and death) because matter tends do disaggregate, so remains a logical deduction , which in those ancient times the brain naturally did not have the same characteristics of a brain today. The biological evolution also came by.

The organic refinement had its progress, and the psychological construction did not stay behind. As the organic organization evolves, the commander of this cellular group, the psyche walks along.

Comparing technology with organic life, the principle is the same. Each refinement, the process of transmission and reception of life also follows the same law.

In short: In order to have an improvement in mental or psychic system, it is necessary to care for the human machine. This is possible through the care directly on maintenance of matter as well in maintaining a healthy spirit.

Moral

Here is an excerpt taken from the book "Great Messages" of Pietro Ubaldi "Message of Forgiveness" where master says:

"Everything is connected in the universe, physical causes and moral effects, moral causes and physical effects. A compression organism evolves you and in it you are trapped in every act".

Using this concept or universal law, one can deduce that an important achievement will be tied to a merit at the same rate, in this way, it would be illogical to reach this state of psychic evolution without working an internal moral.

A wise boss would only give a responsibility to a subordinate if he possessed the conditions for that task.

It would be a waste of energy putting something on unprepared shoulders.

Logically, only the most prepared athlete can reach the podium. In order to advance on the path of super-conscience, morale is one of the necessary attributes.

Morals are synonymous with wisdom, balance and love.

I believe that science has already searched and measured how much energy vibrates and expands when a person is happy. The real happiness is the reflection of a good moral state.

Nutrition

At the moment I am writing this chapter after lunch, as well as the previous one, I feel a hard time exposing the ideas and assemble   all the words, as this is due to sleepiness cause by the action of the digestive system. I can say that I only ate rice, vegetables and some chicken. Quite different from breakfast, which was very light, bread with butter, milk with coffee.

During the morning, my mind was rested and receptive to ideas.

I am reporting my particular for a reason: the importance of food in relation to the mind and the physical body.

It is know that, according to the needs of the body, nutrition must be compatible with these needs.   In this particular case, the machine must be light, where the blood can oxygenate the brain more than working hard on the digestive system.

People who engage in a healthier body, and especially

when the purpose is the use of the mind, usually go for a light meal and choose more vegetable rather than food of animal sources.

In the specific case of intuition or mental elevation, the ideal is a light weight power.

Science itself directs into lighter, varied and balanced meals to keep the body healthy.

Red meat requires greater work of the digestive system, and also their waste remains longer in the body.

For an athlete, this type of nutrition can be ideal, because the focus is physical and not only mind. The necessity varies according to the goal.

This book is not intended to dictate rules to do this or do that, eat this or eat that, the only intention is to give you some guide lines.

Environment

It would be easy to distinguish a party atmosphere with music, joy, dancing and other attractions from an environment of monastery or church or a spiritual retreat.

Which of these environments is correct?

Both are, because both serve the need for which they were determined.

In the case of intuition or an environment for reflection, naturally the second option would be better.

In place of business there are no monasteries or churches or spiritual retreat, so, what to do then?

Generally there are more peaceful or calm areas in the company or institution. Choose this area when there is need, retreat for a few minutes. Forget a little of your surroundings, elevate your mind and heart your own way (there are no fixed rules). Think about your problem at work, or if it was something that just came up, or looking after a client or whatever the need might be. As God or the

Sergio Antonio Meneghetti

Universe to help you to help you or get inspired right now to have the best ideas.

Release your mind; let her go your own way. Soon the answer the answer may come as your thinking. Observe your heart, if it is tranquil about the idea, go ahead, but if the heart is tight, it will probably not be the best option.

Just as an example, on my place of work, I always reserved a spot where there was not a lot of activity. I always did my prayers there, even being in front of some equipment. As a result that place became increasingly positive energized and quiet.

As you can see, we can create an environment suitable for every purpose without necessarily resorting to practices or objects that characterize such an environment.

Remember, it´s all in your mind and heart and not in characteristic artifacts.

At a glance, the more places of peace and positivity there is in a company, the higher the productivity and responses on this.

Just a reminder, the human being is a transformer of energy; he captures the telluric energies and makes them positive or negative. This is not about religion or beliefs but pure science.

Still speaking of environment:

I am going to point out one more interesting fact where intuition was the key to the deal.

At the end of 2005 the company I worked for was moving to Pindamonhangaba and naturally it would be my new employment address. Leaving my family behind in Santo Andre and work in this beautiful place by myself wouldn´t be interesting, so the family agreed to move to that city, but my wife said the following:

We´re going but I want to live in a gated community, because in that city it is cheaper and provides quality of life for us especially for the children.

Agreed, from there we started to look for something that would fit our financial conditions.

We found a good sized corner lot, the deal was practically done. As the realtor had left me with a map of the condo, I looked at other lots. One of them caught my attention and it was in the same street. The difference was that this lot was almost twice the size and of book more expensive.

That stuck in my mind as a very strong thought.

I talked to my wife and told her I would like to buy that lot instead.

She agreed, but with doubts because that first lot cost $45,000.00 and the new one cost $62,000.00 and we didn't even have the money for the first one.

I called the realtor agent and asked about the new lot. He told me:

- Sergio, if you made an offer of $47,000.00 the owner might accept, because they are building in another town and they need the money.

I made the offer and agreed on the condition that I would pay $2,000.00 in back fees and taxes. I still managed to pay it all in three installments.

That was another victory tanks to intuition, because when a strong thought come to mind, it is very important to pay attention.

Even though the lot was large, I decided to build a mezzanine so I would use it in the future as a den and work place (I worked for 35 companies). My wife complained that she wanted a one level house but I insisted and my will prevailed. It´s been 8 years and now I am writing this work in this place I dreamed of. Looking to the right through my window this pleasant afternoon, I see the majestic Serra da Mantiqueira. Each day I watch the sunset as a new canvass.

Nothing came easy for me to get where I am, but it was very important to pay attention to intuition and fight for something that seemed impossible at that time. Today I am my environment.

This is another account to show that everything is first born in the mind, but without faith and work nothing would be possible.

Now a Break to You Friend:

He´s on our side, he enchants.

Our silent friend, our refuge.

Accompanies the creature, without being noticed.

He shows the truths, the lies and what happens.

He cheers up, elevates.

He carries the heart, fascinates.

Creates places in the mind, moistens eyes.

Shows waters and Lunar craters.

Show us a path

Clear as water, or dark as wine.

Opens us the feelings

Or show us the torments

Contains all, of old age to childhood.

Teach us to review, and also tolerance.

It makes us see with transparency

In it we learn patience.

We all owe a lot to his humility

In him contains hatred and also charity.

His fields are most beautiful

And his words form alliances

Still during sleep with us.

Can always be closed or open campaigning.

Follow his information with fervor

Sometimes we get lost, including modesty.

Sergio Antonio Meneghetti

In it we materialize our dreams, countless.

Those who speak of roughness, or lovely things.

Opens our eyes to science

Or numbs the conscience

Can cause a revolution

Be calm as breeze or hurricane force.

Doesn't care what he receives

Can be ice or it could be fever.

The great one of him makes use

The hypocrites of him abuse.

Brings revelation to humanity

It generates excitement, and the eternal light

He´s always there at our will.

Show us the complex and the purest simplicity.

It is the key that the mind opens the gate.

I am talking about the Book that accompanies the whole life.

(Text of the book: World Peace – Volume II)

6

Because

Processing Speed

Inevitable Tool for the Future

Simplifies Processes

Economy of Resources, time and money

Ecologically Correct

Develops the Human Being

Sergio Antonio Meneghetti

Processing Speed

Arguably the world suffers a frenetic movement in all areas, and increasingly the human being has to chase after answers or trying to keep up with the changes generated each time.

It is humanly impossible to assimilate all those changes.

In the case of companies and businesses responsibilities drain human beings in such a way that some professionals live in extreme stress. They depend in their jobs, and at the same time are overwhelmed by the changes.

If the human being only work, he can be outdated, if he works and studies, there is no time for the family or leisure, so the main essence of life that is live and evolve with balance loses its purpose. It creates a civilization outside the normal parameters of equilibrium.

Despite the resources being faster, in many situations there is no time for decision at the same speed of your needs, at this point the instant response of intuition can help decisions or ideas.

Again I call attention to this mechanism, in which the universe moves with wisdom. According to the needs, it will appear the means to track progress.

Parallel to the development of technology, the development of the psyche is very important. This is the path of equilibrium and a future promoter.

Inevitable Tool for the Future

As reported in the previous chapter, the speed of events is a reality, and there is no way to avoid it.

The development of consciousness can only follow the same rhythm.

The path of this development proceeds in a different way. While the technologies and methods are dynamic and in the case of the mind, the tranquility and insight are factors that shape the consciousness. It is the penetration into the very intimate, listening to a voice that speaks like you, like your thoughts. It is at his gathering that the mind dilates.

The knowledge with prejudice, without dogma or rules, is extremely important as food for the mind to grow.

This information leads the professional to meditate the following; it´s no use if it only enriches academically and technically, but also develops internally to be able to change.

Sergio Antonio Meneghetti

As future is inevitable as well as progress, this psychic tool called intuition is a mandatory stop on the tracks of life. It will not be as many methods that "come and go" in the line of events, but it will be essential for everyone.

In the same way that life out of unconsciousness and consciousness, the super-conscience is a mandatory goal.

Simplifies Processes

Intuition by its power to show the phenomenon in a more complete form, can act as a source of new ideas or vision of leaner processes leaner and more productive.

To cite an example, let´s use a parallel to the radio.

The vast majority knows vinyl recording. In order to put information on the disc there was necessary to compress at a maximum the spaces.

There were developed the best needles to pay songs and also equipment to work the sound by giving them higher purity. There was a big job and various actions were required to simple produce a purer sound.

At this point the technology arrived practically to its limit or saturation point. When something reaches a saturation point, a new cycle starts, so it is in life, as in all things and in the universe.

In the case of sound information record vinyl, someone had the great idea to generate the CD.

This new form of technology revolutionized the industry in this area, and it has open space for other areas such as information technology, for example.

This is an example of the leap from idea, if the search would continue only in that sense; it would reach a point of no exit. The new always comes to meet the needs.

In short, various equipment and peripherals to produce a sound of good quality have been synthesized and with better results.

Curiosity: this will also be the future of writing and it is already in progress. Write less transmitting more.

Many use complex language to explain the simple.

I prefer to use plain language to explain the complex.

(For the word to be universal and not only intellectual).

(Text of work: Freedom of conscience)

Economy of Resources, Time and Money

Let´s create the following situation:

A company imports a product "X" of another group unit in Europe to meet a demand in the automotive industry.

A technical question:

The possibility of producing this same material in the country.

The company receives the following response from the staff:

To produce this product in Brazil, you will have to import a specific plant resin from England and set up a system of million dollars to work on the final product.

According to calculation to assemble this structure as a whole, has made the project unfeasible.

In short, through intuition this product is produce in Brazil regardless of specific resin, and without building a structure of 6 million dollars. The fact already answers regarding the associated knowledge, intuition can generate profit and saving time, resources and money. (See: Low Odor ahead)

Ecologically Correct

Having acted for a long time in the area of product development, I will use example correlated .It is normal for people to notice the rise of plastic components in cars.

Many people, for ignoring the subject, may find that this is only to cheapen the vehicle or give an aesthetic touch. They are right about these two items, however, something deeper in this case directly impacts people´s lives.

Plastic parts in vehicles are aesthetically and more aerodynamic. Also reduce the weight of the vehicle, avoiding consumption of mineral resources and can be recycled.

Not entering the complexity of production, a plastic part can be produced in seconds, while a metal takes longer, requires paint and other chemical treatments, thus creating unwanted waste and consumption of resources such as water.

A lighter automobile with better aerodynamics consumes less fuel and generates less waste pollutants. As automobile production is so large, thousands of pollutants are no longer released to the environment.

When someone develops a plastic resin with greater rigidity, lower density, higher impact and lower viscosity, this person is contributing positively to the environment.

This is another example where the intuition helped and is listed as an intuitive casa further ahead.

Speaking or economy, just remember the nature.

Here´s another example of how important is paying attention differentiated and stronger thoughts.

Try to be aware of this type of mental reaction, because these are the subtle messages that get great ideas. Try to observe from this study how the source will begin to flow from your inner self.

Let´s go!

On the evening of March 15, 2016 I was home alone when there was a strong desire to work on another book I had started years ago.

It was about a water molecule reporting to its passage into the terrestrial ORB. Imagine the water telling her story.

The inspiration was so strong and continuous and also surprising that in 6 days that little book was finished.

As I finished the book, I felt sure that, once again, to believe in the tip of the ideas and give the same flow. It is like seeing the tip of an iceberg and then dive in and understand its dimension.

Except that, something more interesting came up. On that day, March 21, when I finished writing, a thought came to my mind:

When is the "Water Day"!   This idea came to my mind several times several times until I decided to look it up on the internet and to my amazement; Water Day was the very next day, March 22.

It was early evening already, and with a great effort I composed the book cove and the publication process and on March 22 the book was published: "Water Life".

I must confess that in certain parts of the story I was moved,   thus along with the story, shedding some water from my eyes.

Note that in the cases mentioned, I wasn´t preparing to write the books, but simple gave vent to these perception that arrived light a lightning bolt to my mind. Then I repeat once again: pay attention to your thoughts.

Just an observation: Due to the need and cost of publishing a book, I started to walk the hard road to learn how to do this process.

Of my 14 works published, the first three I paid the other 11 I did from start to finish, including work I have done for 8 other writers.   Today, the dreams of these talented people were published for a low cost.

This book that you are enjoying now, was also the result of a great idea that is, earn your wages teaching what is very important to me.   And I know the importance of this subject "the perception and intuition development" as a tool for the progress of making.

Sergio Antonio Meneghetti

Develops the Human Being

For everything that has been previously written, one can deduce that the expansion of consciousness quietly moving to super-consciousness, develops that human being as a whole, both in the material area and spiritually.

Growth is the primary factor when someone chooses this path.

Intuition opens a vast range of choices in people´s lives and progress.

Academic development is very important, but is insufficient to shape this part in transformation, which is the human being.

The living being has its contacts with the external environment, registers, absorbs, works the information and after contributes its share to progress of the whole, so the cycle goes on.

As the trend of progress is the refinement of everything evolves, and human beings also suffer this phenomenon going on the intended. While there is need for the use of this demonstration, the universe will be providing this assistance.

Intuition develops the human being, giving him conditions to understand phenomenon or information outside time and space.

This intuitive path leads the individual and professional to the most advance form of research that exists; the full perception of the phenomenon.

It would be vital to scientists and researchers to have consciousness of this fact, and the use for the advancement of science.

A note:

I believe that during this reading you mind started looking for situations.

And these situations you started to connect to the subject of intuition.

Your mind is working and progressing already.

Well thought = construction.

7

Company

Definition

Work

Professional Features

Professional Environment

Spirituality

Human Resources

Results

Sergio Antonio Meneghetti

Definition

It is normal to look at a company as a place that produce goods and provides services.

Um place where the owner needs other people to perform a job and them an amount for carrying it out.

In fact, a company is much more than that.

A company besides producing goods and earn its livelihood, collaborates with its grain of labor on the large construction of progress.

The company is an environment where differences are tune in with a single goal, and this coexistence generates feelings such as tolerance, solidarity, complicity among other virtues.

A company is an environment of human refinement.

It is a blessed place because it enables human growth.

Sergio Antonio Meneghetti

Work

Work, contrary to what many people think, is workshop where the human masses spend their energy generating human progress on the planet and on their individual lives.

I will quote again another excerpt from the work of Pietro Ubaldi. Great Messages-message of Forgiveness:

Love the work, including the work material.

High and Holy thing...

Love the work, but with new spirit, love it not for what it is exactly, but as an act of worship to God, as a manifestation of your soul, never as wealth fever or domain. Do not attach your soul to the results, which belong to the matter and therefore subject to inspire love, the act alone, and the act of working. Let it not be the possession, the triumph your reward, but yes, the intimate satisfaction of a job done, each day, each day your duty, so collaborating to the functioning of the great organism.

It may seem utopia or far from daily reality, but as all

moves in that direction, one day the professional will have that goal and the result as many preach: "do your best and the result will be a consequence".

Nature works in harmony, the dynamic moves everything in the universe.

I dare to say that: the function of work and company will suffer transformation in their philosophical bases.

Professional Features

As a subordinate or employee, imagine the characteristics of an ideal boss.

Now imagine yourself as a boss, and under your command the ideal professional.

This is the best formula to gauge yourself in the professional field. Only putting yourself in the place of another, your may have a better idea of the scenario.

It is worth remembering that the goal is perfection and not the moment. This is an eternal construction, so never expect perfection from others as we are still imperfect.

Even though that the professional thinking who wins more with his good performance is the company, in fact, he is the greatest winner. Nobody loses for doing and seeking to be the best.

(See the text: The new worker in the work, intuition work tool-author).

Sergio Antonio Meneghetti

Professional Environment

Needless to say that the success of an orchestra is allied to the maestro, the pitch and the order of the musicians.

What is an orchestra but a well-managed company to produce a great product?

To know how to organize the set is not an easy task, but without a direction and harmony an execution would be impractical.

On the other hand, if the musicians didn't observe and followed the maestro, hardly the results would be valuable.

The focus of this chapter is just the "harmony".

Who´s to benefit from it?

The orchestra?

The musicians?

Everybody wins. The victory is the result of the whole group.

Sergio Antonio Meneghetti

When it comes to science, the first idea is to research, atoms, studies, physics, chemistry, mathematics, etc.

In a more profound vision, relationships respect scientific and mathematical laws. The addiction and subtractions in the end results are facts, they will always have behaviors that as in chemistry will generate good or bad products, and by-products also. In physics, the law of "action and reaction" is indisputable fact.

In an enterprise, such in an orchestra, all are within a crucible of purification. The effort and suffering of each element by the warmth of human relations will cleanse and generate the precious metal.

It´s good to come home after a day´s work, and next day not felling bitter in return to this work.

The greater the harmony, the greater is the possibility of intuition to act.

Spirituality

Always keep in mind: your job is your job, never mix things.

The human being sees in work or in other areas, distinct function and without any link, for example: Science and Religion, for many se opposites. But is there something without any link in this universe?

Everything is connected somehow, because we live in a universe where a mechanical thing complements the other, and only that way the gears can work as a whole.

The work is not disconnected from other areas, such as religion or philosophy.

If I cultivated a discrete prayer in the professional environment, of book I will be moving positive energies, that will favor my person bringing equilibrium, security and clarity in ideas and decision, and it will also be generating positivity in the environment.

When it comes to energy, they are detectable energies, not fanciful things or religious or magical. It´s pure science.

I am sure of this fact; I can have it as a philosophy of life in the professional field.

Returning the first sentence, work is work, so having an attitude of prayer (no matter he creed) it should always be discreet and not to turn the workplace into a church.

Human Resources

The art of managing people. That´s a great challenge.[

The wise difference between to order and to lead. One requires strength and hierarchy, the other, morale and hierarchy.

The first attitude hinders individual freedom, the second releases through knowledge.

I believe that to lead people is one of the biggest responsibilities that human beings can have.

To provide a way to the future of fellowman is to provide a way to the progress of the planet, or to obstruct the same.

How much suffering is generated by the lack of wisdom on human behavior, or how much benefit can be generated when it distributes knowledge?

Equation

Sergio Antonio Meneghetti

The company success is in this equation

For all, law and responsibility

These are the four and the yeast for the bread

Bread for all, synonymous with equality.

When a company reaches this point of equilibrium, the same is realizing its real purpose before the world.

Growth and justice to all corporate body is the wisdom in action, it is the collective refinement walking in the same direction, and harvesting the fruits previously sown.

It creates not only the material but also it wins peace as a whole. And this peace will become a safe haven for new advances in the future.

(Text taken from work: Management is an Art)

Results

There is unanimity among employers and employees in the following context: make the most performing the least.

There is no mistake about it, because it is made an analysis in nature,   this science would be used. The use of only the necessary in order to produce maximum results without waste.

Insects as other animals build their shelter according to use (just as an example).

The collective welfare is the main goal to be achieved, because a parent always seeks equal distribution for his family.

Observe that to get to good results, there were created a line of ascension on the concepts and practices.

These same concepts and practices are as in the attitudes and actions of the body of a company, institution or society. And this set up has a goal, to give conditions for human beings to expand their awareness and build super-conscience.

At this point, a civilization or nucleus may be able to give went to their inner potential where they are housed, in silence, the greatest treasures.

Internal results are eternal, while external achievements are fleeting as time.

In short: the construction of the thought or the construction of the human being is the best results that can be achieved.

At this time

I asked God with Love and Unity

You reader, be a thousand or a million

It´s the least to whom will help me earn the bread.

This is what I can do; as desire for gratitude.

8

Real Cases Experienced

Security

Fire

Hose

Assault

Glasses

Achievements

Japan Projects

Volker Trautz

Barack Obama

Low Odor

Sergio Antonio Meneghetti

Decision Making

Sale of Property

Invitation to Society

Sale of Plastics

Science

Sea of energies (Higgs Field)

Formation of matter

Center of the galaxy

Curvature of the universe

Birth of Hydrogen

Miscellaneous Cases

Security

Case 01 - I work in a company for six years, all that time in relay arrangements, i.e. I worked seven days a weeks in each period, morning, afternoon, night.

On July 24, 1982 I was to start a shift from 4pm to midnight. Around 2 pm I was having lunch when a very strong thought came to my mind, as if someone spoke in my ear:

Fire at the factory!

And this thought came with some sort of flash.

That left me anxious and worried.

I went to work that afternoon with a restless heard.

Around 5 pm I heard a small pop, looked at the outside of the lab where I worked, and toward the area of production and noticed a leak on the side of a filter. This leak launched a continuous jet stream, but without greater danger.

This small problem gave me a sense of relief, because I figure it would be this the resulting phenomenon that felt earlier, but around 6 pm I heard another pop, accompanied by a continuous loud noise. I went out by the side of the lab then I noticed large flames resembling a torch. It was the fire announced at lunch time.

That intuition was important because it alerted me to the danger that was coming.

There were fire prevention actions taken by the company, of which I was part.

Thank God, there were no victims but the fire lasted from 6 to 8 pm.

For this type of intuitive event there is no control. It does not depend on our will, to work in order to be more receptive we can.

Case 02 - In 1996 I participated in a project call "Multi Skill". The scope was to multiply our knowledge in other areas of the company, so I was transferred temporarily from the lab to train in petrochemical process.

There was a routine for loading and unloading wagons of Propylene to a reservoir (sphere). This maneuver required a lot of security, there were two block valves, being a manual valve and another automated with a distance of about 40 cm between them.

In the end there was a steel hose wrapped with a plastic coating (a flexible tube) with a diameter of about two inches, which would be connected to the cart.

As it was the rule to test everything to avoid accidents, I opened the automatic valve to verify that there was no gas between the two because the gas pressure was high, eight atmospheres. At the moment I was to rotate the valve, something came to my mind as if someone was talking to me:

Get out!

I felt great apprehension, moved back as far as possible and turned the valve. The hose attached to it jerked like a whip, right in front of my face with such a velocity, throwing gas. It hit the floor with such a force that it chipped the concrete.

If I wasn't attentive to intuition, I would be dead, because I would be in a place within the range of the whiplash.

Case 03 – Around the year 2000, I was one of the partners of a small plastic company.

One evening a meeting was schedule with the other partners. We were to talk about the possibility of starting another business of recycling waste from the grounding of plastic bottles.

On my way the meeting, from my home, I started having an uneasy feeling mixed with fear and anxiety. I changed the way I usually drove, but there was a little section between the city of Santo Andre and Maua that I could not avoid.

At that point, around 8 pm I was blocked by a VW Buss, a young guy jumped out and pointing a gun to my face ordered me to move over to the passage seat and not to look at their faces.

Result: They took my car and up ahead in the outskirts of the city their let me off unharmed.

Case 4 - A certain night, I am my wife went to pay a visit to a friend of mine from the   Academy of Letters of Pindamonhangaba, Beth Guimarães.

We talked excitedly, but I was unease about my glasses. Fixing on my face, taking it off and into my pocket, back to the face, and this was becoming quite uncomfortable.

During the visit, a quick and heavy rain came down.

After the visit, it was raining a little still, so I ran to my car so not to get wet.

At the moment, my friend noticed a peace of branch on the windshield of my car, so she gently came outside and removed it for me.

In an automatic gesture, I reached for my pocket to get my glasses but realized they were not there; I looked quickly in the car and nothing. I opened the door and saw it on the sidewalk, wet and broken.

That´s what happened: When I ran to the car the glasses fell on the sidewalk, and when Beth was removing the branch from the windshield she stepped on them.

Observe that the intuition serves on various levels of cases, i.e. It is important to pay attention when something inside of us is manifested with different intensity.

Achievements

Case 01 - I worked in the area of product development in the plastic business, being   national automakers our biggest customers.

In the book of the years 2000, the company had some ongoing project in order to come up with three new materials.   These materials were designed to meet the needs of a Japanese automaker. In order to facilitate the project, it was decided to import raw material and formulations from a Japan-based group company.

With formulations and material in hands, I started a pilot production in our lab.

After finalizing the production the samples, there was the characterization of the same.   In order to verify the quality of the products tests were made according to requirements of the client.

It is worth to note that the same product have been successfully manufactured in Japan using the same raw materials.

It was observed that the products here in Brazil were not of the qualities and thus would not be approved.

These products would have to be tested in Japan.

There were new productions and again without success.

A question arouses:

Send the products just the same?

As it was Friday already, I asked my boss if I could make another try on Saturday. He gave me the green light, even though he did not believe I would have success due to the results of the previous attempts.

Next day I was there for that challenge.

What would be the first step, since I had the feeling that I was going to succeed?

The "how" I did not know.

In the tranquility of the sector, because it was a weekend, I started to look at what we had done. Of book it wouldn´t be the right way to go.

It would have to be something new.

So, I started to do what do when something is very difficult and it is beyond my understanding, I began to pray asking for help from heavens.

A possible key to the goal began to draw in my mind, showing a new way. Now, left to make a rational analysis of intuition and what to do to make it happen.

After an analysis of the subject, I started the process. Did the production of compounds and the prepared injection of body proof) for testing on Monday.

Finally Monday arrived. I started the tests, and in the first phase, at 23 degrees Celsius, it was possible to see a great improvement.

I set aside some material to be tested at low temperature. That would be tested after 2 pm.

A little before 2 pm my boss came by the lab and asked me if I worked Saturday. Then he asked me if I had any test ready.

Even though having the first phase ok, I waited to complete all the other tests before announcing any results and told him I wasn´t ready.

Again he made a comment to me and a co-worker next to me:

> - Sergio! I understand your effort, but I sorry to say that it is not going to work.

I didn't answered, but we knew already about the positive results from the first phase, and by experience, the next phase probably would be positive as well.

At the right time, we finished the tests. As anxiously expected the results were excellent.

Next day, there was lots of praise from our bosses. It´s worth pointing out that many victories have been achieved thanks to the autonomy that we were given, and without that autonomy and support much work would be stalled.

Result of the work: there were sample redone and sent to Japan. The results found in the friendly country were better than expected.

After this testing phase there was another commercial dilemma: production costs were very close to the price of imported materials due to imports.

After meeting with the client, it was suggested a national version that met the minimum requirements of specification, but at a lower cost and with mainly national materials.

Based on new processes, technology obtained through the intuitive process, the team began to develop the new versions, and in a short time we managed products with properties that best served the client´s expectations.

In short: we achieved the goal and we started selling the products.

These new products stood out in the automotive industry attracting new clients.

Another important step: the transfer of this new technology to all products with similar characteristics, obtaining in this way a reduction in use of imported raw material, thanks to the increase of this properties.

What at first was just a need for a project, ended up becoming a principle of economics, leaving several products with competitive costs, generating millions in profits overtime.

With this new technology the materials have become increasingly ecologically correct, as with better properties they generate lighter parts and as a consequence less fuel consumption, less wear and less pollution.

In this case, intuition collaborated beyond the expected, shortened research and testing, saved money, time, manpower, raw materials, prevented waste from tests and showed a new technology path.

This achievement could have come across anyone using

the intuition as a working tool that is why my credit in history is just to use the tool at the right time with a competent team.

Observe the potential that each professional can develop in his own work, as well in their personal or artistic fields.

The universe offers us an inexhaustible source of valuable information, it is up to us to discover how to open this door and best channel the best of it for the betterment of the whole.

This is a direct manifestation, that is, use this intuition wisely to achieve noble goals, even if they are considered small goals.

Always be aware of the thoughts, as the mind might be gathering information for an important task ahead.

Case 02 – I have a friend that ran an international company, he even had a job abroad.

Due to several factors we talked little, sometimes we exchanged emails.

At that time, I felt an enormous need to communicate with him. I had a hunch that I should tell him about a diversification in business of the company he ran.

That thought was already bothering me, I should take some action, repeatedly I rehearsed, sent him emails, but at the same time I thought it could be silly on my part, and also, of book, he would know what´s best for the profile of his company in the market. He was the most qualified person for this.

Sergio Antonio Meneghetti

That thought would not leave my mind, and finally I mastered enough courage to send him my intuitive opinion.

To my surprise, he immediately answered my email in these terms:

Thank you very much for the information and referral. We are involved in a detailed study of this business also in Brazil. We are awaiting the results of the second quarter of 2007 to decide later.   Anyway, I appreciate your tip, thank you very much.

I have never had any contact in which this type of subject was mentioned because it was something confidential.

Now, remain the following questions:

Why such strong intuitions to the point of bothering me like this?

How would a subject come to my mind that was already under way in another country?

In my humble opinion, probably it is a warning or positive reinforcement for the subject in question that it possibly would succeed.

It is up to the time for this subject to mature and bring up results, a results that I hope will be the best possible.

For ethical reason and of privacy of information, I cannot give any more details about that business.

Note:

(Text extracted from the book: Intuition, Work Tool).

This is the second edition of the work, and below is the record of the time when I wrote the work:

Obs. Today, December 21, 2007 I leave registered the creation of LyondellBasell Industries, the third-largest chemical company on the planet, with CEO Mr. Volker Trautz author of the email above.

Case 03 – After the tragedy of the earthquake in Haiti, I wrote an article on the subject, seeking to show some values that deserve attention. The same had great repercussion in the media of Pindamonhangaba.

One afternoon, while performing some test on the job, I had the urge to send this article out of Brazil. As I was working, I tried to let it go, but the feeling became stronger and even annoying.

I went to the computer, did a quick translation via Google and set to US State Department. There was the traditional automatic response thanks for the message.

I went back to my tests, I continued with the same feelings in my heart.

I remembered seeing something about Barack Obama´s blog; I came to the computer and searched the site of the White House, until reaching the contact of the President. I filled out the requirements; I added the translation via Google and sent it. The web site responded "excess". I decreased the amount of characters and sent it again. It returned the same answer. I decreased it again and finally there was a positive response of thanks returned.

Good Friday week, in the beginning of April 2010, I had a visit from relatives in my home and that morning I made a comment to my older sister:

- Nilza, I sent a message to the White House and it is a

little quiet, I think they might be paying attention.

In the afternoon, checking my emails, there was a big surprise: There was an email of thanks (Thank you for your message) I opened the email and it was a presidential message from Barack Obama thanking me for the article.

This is the email:

Thank you for your message

From The White House-Presidential Correspondence (norply-WHPC@whitehouse.gov)

Re: You may not know this sender. Mark as trusted / Mark as junk

Submitted: Friday, April 2, 2010 18:13:36

To: Sergio.xxxxxxxx@hotmail.com

Dear Friend:

Thank you for writing regarding the situation in Haiti. The earthquake that struck Haiti on January 12 shocked the world. The loss of life is heartbreaking, and the suffering are devastating. The images of this tragedy remind us of our common humanity and have invoked our Nation's enduring spirit of generosity and compassion.

My Administration has responded with a swift, coordinated, and aggressive relief effort, among the largest in our history. (I) designated Dr. Rajiv Shah, Administrator of the United States Agency for International Development, our government's unified disaster coordinator. He is leading America's effort alongside the United Nations, together with international aid and nongovernmental organizations on the ground in Haiti. I have also enlisted the help of

Presidents Bush and Clinton, who have launched a major fundraising effort for Haiti, and those who wish to help should visit: ClintoBushHaitiFund.org.

With the pledge of our full support, I assured Haitian President René Préval that America stands by the Haitian people. We must meet their needs through sustained assistance to help Haiti recover and rebuild. Bringing relief to the millions who are suffering poses tremendous challenges—navigating crumbled roads and damaged ports, ad finding shelter for the homeless—but we must forge ahead to help restore the Haitian people´s energy and optimism for a more hopeful future.

We are fortunate that our Nation has a unique capacity to reach out Mrs swiftly and broadly, and Americans have always eats together to serve other in times of great need. The dedication of our military personnel and rescue teams, and the goodwill of millions of Americans lending a helping hand, demonstrate the courage and decency of our people.

To learn more about our efforts, visit:

www.WhiteHouse.gov/HaitiEarthquake. We will stand with the people of Haiti and keep them in our thoughts and prayers.

Sincerely,

Barack Obama

To be part of our agenda for change, join us at www.WhiteHouse.gov

Following article sent about Haiti:

The Renaissance of Haiti

"Old seeds that build the new garden"

Haiti, a land marked in the beginning with slavery,

For the past little distant by war and abuse,

Today marked by nature with its force of destruction.

Tomorrow will be reborn with love and construction.

On the tracks of life man passes through several stations, these being necessary for their achievements, freedom and maturing. This is his eternal walk.

The rocky movement in the recesses of the Earth, practical action of nature without discriminating anyone or their places.

Now mingle rich and poor, ignorant and scholars, commanders and subordinates, politics from the right and left, old and young, believers and atheists, men and women, strong and weak, friends and foes, black and white, proud and meek, idle and workers, thugs and victims, bosses and employees, donors and recipients, so the duality lose what is ephemeral, only remains the result of the work of each.

While the world searches for victims and solidarity, **charity and compassion** are present, those that stay must rebuild a nation, however, that this may be its essence,

PEACE without distinction.

That the elder brothers of the globe may do their portion of kindness, helping with material and knowledge. That they may help turn this place of suffering into a new one where prosperity may be dynamic and productive.

Teach this people to extract from their work growth. Almsgiving is welcome but quickly drains and is lost in time.

To the people, when burying their dead, bury the negative past too to keep only its qualities and new perspective for a better life.

Remember those obstinate people in doing good that left their lives in Haitian soil. They were many and they all deserve our gratitude.

The pain, this thing that reaches us all showing how human beings are frail and what really has value in this life. It arrives in silence, sometimes tame, other times fierce showing all tis power, it is bad to our sight, but it is also a blessing that awaken us from illusions showing us the good and straight path (we only lack understanding towards it).

Pain corrodes feelings of love and solidarity in the heart of others, stirs emotions and thus the world softens and a yarn for improvement is present. This is the opposite of violence that generates friction and feelings of revenge.

History has shown that many countries that suffered natural disasters and war destruction raised and became better than before (all thanks to the strength of the will and action of its people).

Haiti is the room in this great house that requires attention and reform; we cannot forget that there are other Haities that need help and attention.

Sergio Antonio Meneghetti

For each coin placed to the progress of our brothers in need, I am sure this will be the best dividend in the future, namely, the gratitude of having taken a step beyond ourselves.

What is the biggest scope of life or happiness?

That each contributes is quota for its construction.

I leave my thanks to all the heroes of this venture that somehow contributed, contribute and will contribute rebuilding this Haitian home.

Sergio Antonio Meneghetti   01-18-2010

Between article submission and the response from the White House a phenomenon occurred:

In my mind, I saw several times, a huge hand holding by the fingertips (indicator and thumb) a small piece of paper.

In short:

What are the odds of someone from a rather small town in Brazil, not famous or from Government, send a message (translate by Google) to the President of the richest nation on the planet and get a personal reply?

I think it is easier to win the lottery.

This is another fact that show the potential of this tool called intuition, what seemed absurd turned reality.

Case 04 - Low Odor

Almost everyone like the smell of new and in the case of an automobile that is more pronounced.

For reasons that don´t relate to this case, the auto industry felt the need of reducing the smell from the internal components of the vehicles.

As usually, the specs are created in the headquarters of the automaker, and usually those are in foreign countries, the first products to meet those requirements in this aspect were also created out of the country.

In order to meet certain European manufacturer, the company where I worked, imported a material with specifications to meet the new requirement, "low odor".

Importing any product has several low points, from pricing, logistics, storage cost, so the company where I worked, technically looked into the possibility of producing this product internally. The answer was the following: to produce this type of product it would be necessary to import a resin "Z" of a specific unit in England, and also set up a post-production process at a cost of around 6 million Dollars.

The cost for this purpose and a not so expressive consumption internally, made this project unfeasible.

As there were victories as in case 01 in this chapter, I dared to do the same with this type of product also.

When we want something honestly for a good purpose, things start to come to us.

In conversation with a supplier, the first step came up, and with the help of intuition was born the second important step.

Attending to what was intuitive, I did the same that is, I followed what came to my mind and acted in the process of testing.

The result was positive. Days later I followed tests in the plant on two types of equipment, and the results remained positive too.

There was another important achievement for the company and for the customers, because it would be more secure to produce and distribute a single product internally rather than go through all the troubles of importing.

The important thing was generated technology and the costs were also agreeable and competitive.

Once more, intuition or psychic process was decisive in the progress of the company.

Decision Making

Case 01 - At the beginning of 2010 I was to make a real estate deal, and if this wasn't done quickly I run the risk of losing two properties.

My property in Santo Andre was mortgaged to raise funds for a larger building, and selling that to finish the construction of my new house in Pindamonhangaba and pay of the existing mortgage.

I had to make an urgent decision that would solve the situation.

There was an alternative, however it was below the value, so I sought to get informed to make a decision, but without success. So, my wife said:

Well, didn´t you write a book about intuition? Use your intuition!

That's what I did.

I prepared during the week taking care of the emotional and spiritual balance. I separated the situation two ways:

a) I visualized myself making the proposed deal (the sale of my St. André house), that would be the immediate salvation, for a 30% below market value, though.

b) I visualized myself not accepting the deal and waiting for a better offer, even at risk.

I paid attention to my heart, namely, the situation in which I would feel a quieter heart should be the option to follow.

Result: my heart felt good not closing the deal, though I may lose everything, it was a big risk.

So, I did not accept the offer on my house. Thanks to God and the help of intuition, another offer was already underway. I accepted that second offer that was much better.

I was a good decision, in which, without the intuition I would have I would have listened to reason and in the end I would have made a bad deal.

Case 02 - On the evening of April 12, 1999 (my birthday) I dreamed that four people were inviting me to be partner in a small business. The four were with sad faces. The next morning around 10 am, a friend from the company came to me and offered me partnership along with three other partners, immediately I remembered my dream and felt it as a warning, but I did not pay attention.

Result: I joined the partnership as a fifth partner.

In this company there were only large monetary losses to me and lots of suffering. That would explain why the sad faces. It is important to pay attention to the details of intuition.

The only good thing about that time in that partnership was the experience and courage gained and during that period I wrote my first poem "Writing". This poetry opened the doors to what I do today. Even in suffering we can overcome things, reinforcing that theory that suffering in the spring that pushes us to personal growth.

The case came through a dream.

Case 03 – For some time, my wife was in the plastic business, selling recycled material of good quality.

Once she had to make a sale of good material for an unknown client.

It's worth noting that in this parallel market there are good people, but also less honest ones too.

My wife called me and asked me if I could deal with this person, since he would take all of the material, of considerable value and he was paying by check which was kind of risky.

I did what I normally do, and accepted his check, trusted him.

The client got his material the check went through.

A similar case happened in the purchase of a high valued material that would come from an unknown supplier. The payment was to be on delivery and the supplier was from a large city. Dealing with plastic scrap, problems can arise such as contamination by mixing, cleaning or exchange of material. Guided by intuition, the material was purchased and the quality of it was within the expectations. The result was positive.

The importance of decision making is in proportion to the impact that it will generate.

In a the case of a large company, it can involve millions or even the future of the company, in the case of a family or personal, the impact is less apparently, but It can mean a lot to someone.

Science

After the cases mentioned that I personally had the opportunity to experience, we then come to a very specific section in which we have the opportunity to show real cases related to the theme of the book: science. Such cases are important to present intuition as an important complementary tool to support research and direct its paths without straying from rigorous methodology.

Case 01 - When I was writing the book "The Reconstruction of the Universe" in Orlando, in 2015, I had many doubts about how something could expand in "nothingness".

I say this because, according to the Big Bang Theory, a point would project itself into space, forming matter, generating volume, even if it was tiny.

While I was questioning myself, I lay down, looking at the ceiling of my room. Suddenly, the following question came to my mind in the form of a stronger thought:

Imagine if you removed the ceiling and walls of your room, imagine removing all the objects, including all the stars in the sky. Now imagine an infinite void. Now imagine that this infinite void is a sea of dynamic energies so tiny that no instrument can measure them. This is the energy storehouse that forms matter and everything you see.

At that moment, I understood that there is no "nothingness", but something that escapes our rational and analytical understanding. That the Spatial Dimension is only possible through the condensation of these energies. Thus, this dimension can only exist with the presence of matter.

Case 02 - One afternoon, between 2017 and 2018, I was in front of my computer about to start my work. I wasn't thinking about the subject.

Suddenly, to my left, as if projected holographically, the image of a kind of metal the size of a peach pit appeared. The color was similar to bronze, and, in the center, it looked like a reddish bronze. Near the lower end, there was a wave with sharp points the color of incandescent volcanic lava. As the wave adhered to the object, its incandescence gradually faded, and it became part of that object. This phenomenon took a few seconds, but I watched it in slow motion.

At the same time that the phenomenon was visualized, the information came in the form of a thought:

This is how matter is formed!

I must emphasize that this vision was not how we naturally see things, nor was it a thought. It was like "seeing with the mind."

Analyzing the phenomenon and describing the process and the information that the vision wanted to convey, the conclusions I reached were the following:

- The object, in fact, was rotating at a very high speed

around its axis;

- The sharp spines show that the object was formed by waves of extremely high energy and speed;

- Adhesion: it is like air molecules that adhere to a tornado, integrating it and increasing its size and potential;

- The bronze color expresses the speed of the coiled waves.

Let's look at the physical logic of the phenomenon.

The concentration of energy (waves) could only happen as a kind of ball of wool being wound up. One of the characteristics of the wave is its linear projection, as occurs with the spiral cables of old telephones. The best way to package a line is to wind it around its own axis in a ball of wool, resulting in a sphere, a particle. This is the only way to concentrate a high amount of energy at one point.

Case 03 - Do you remember when I was shown the Formation of Matter? Minutes later, that vision was followed by another.

Beside me, a gaseous disk two meters in diameter and thirty centimeters thick with an empty center formed. It rotated slowly, and I knew it was a galaxy. Seconds later, I saw myself in its center – its empty part – and observed light waves descending along dark vortical flows. It was like a drain with water dragging luminous threads in its vortex.

The information and understanding of the process came instantly.

Case 04 – On the morning of June 19, 2018, while I was drying my hands in the bathroom, a clear image of the universe came to my mind: a spiral full of galaxies – and the idea that light followed the flow of this curvature.

In other words, the light would follow the curvature of the arms of this spiral-shaped construction.

I paid attention to this phenomenon because it was not a common thought and because I was not thinking about the subject. It came quickly and with the information, or understanding, of what was happening. This is one of the characteristics of Intuition.

Simplifying the information, light, over great distances, follows the fabric of the universe or the curvatures that we cannot detect or perceive due to an observation from within the whole. It would be similar to a bacterium wanting to describe the shape of its host. I believe that the new James Webb telescope, launched on December 25, 2021, will be able to show the universe with greater amplitude, surpassing the famous 13.8 billion years estimated by the Big Bang Theory, and, thus, visualize what was described above (vortices composed of galaxies).

Case 05 – The Birth of Hydrogen and Atomic Growth. Since October 2023, I had imagined how to describe the physical process of the transmutation of the atomic nucleus to form the atom, but I confess that I did not know how this happened.

Deep down, I had the intuition that I would discover what its path would be and how this growth happens in the intimacy of the atom.

On December 19, 2023, I had the pleasant intuitive

experience of how this mechanism works. The information came quickly and surprisingly, as it was outside of my analytical deductions while observing the five atoms presented above.

The information came in the form of a thought (the term insight can be used in this case): the electron that orbits hydrogen came from the neutron. After this information, I quickly checked the comparison of the elements hydrogen, deuterium, and tritium. The idea began to make sense, and, soon after, I compared it with helium-3 and helium. To my delight, the idea fit together perfectly like pieces of a puzzle.

I was so moved that I cried with happiness. There was the key to my questions. And although the idea may seem absurd when we first look at it, thanks to intuition and analytical work, it makes perfect sense.

Detailed information on these scientific cases can be found in the following works:

- Do Hidrogênio ao Hélio – Sem Fusão Nuclear

(Portuguese version);

- The Quantum World and the Expansion of the Universe – Cosmological Model by Vortices

(English version written in partnership with an important scientist from Kansas State University – Dr. Lior Shamir).

For those who like science, these two books are revolutionary.

Sergio Antonio Meneghetti

Miscellaneous Cases

Case 01 - US Navy Program to Study How Troops Use Intuition

BY CHANNING JOSEPH

MARCH 27, 2012 5:09 pm March 27, 2012 5:09 pm 5

The US Navy has begun a program to investigate how service members can be trained to improve their "sixth sense," or intuitive ability, during combat and other missions.

The idea for the project comes in large part from the testimony of troops in Iraq and Afghanistan who reported an inexplicable sense of danger just before encountering an enemy attack or colliding with an improvised explosive device, Navy scientists said.

(Credits: The New York Times – www.nytimes.com)

Case 02 - Can we trust our intuition?

Laura Kutsch on August 15, 2019

As the world becomes more complex, making decisions becomes more difficult. Is it better to rely on careful analysis or trust your gut?

"I follow my intuitions," says investor Judith Williams. Of course, you might think, "me too," - if the choice is between chocolate and vanilla ice cream. But Williams is dealing with real money in the high five and six figures.

Williams is one of the lions on The Lions' Den, a German television show similar to Shark Tank. She and other participants invest their own money in business ideas submitted by competitors. She's not the only one who trusts her gut. Intuition, it seems, is at an all-time high: Bookstores are full of guides advising us on how to heal, eat or invest intuitively. They promise to release our inner wisdom and strengths we don't yet know we have.

(Credits: Scientific American – www.scientificamerican.com)

Case 03 - *Intuition of a teacher*

Teachers rely on it, often dozens of times, every day in their classrooms. Sometimes it works; Sometimes no. It usually makes it easier to trust experience.

BY LORY HOUGH

The class she taught at Larsen G08 had just ended and Kitty Boles, Ed.D.'91, then a senior teacher, was packing up her materials when a 20-year-old approached her and said he was preparing to be a teacher of physical. The young man also told Boles that she had been his third-grade teacher in Brooklyn 15 years earlier and that he remembered, more than anything else, something that made a big difference: her hugs.

Boles remembered the boy and how his family had just moved to the region from Iran. She remembered that he didn't speak English and was scared to death. She also remembered hugging

128

him.

"It was intuitive for me to hug him," she says, "to make him feel safe and loved." - Kitty Boles

Although Boles was a scared new teacher at the time, she "felt" what her student needed at that moment, not based on test scores or scientific data given to her by an administrator, but based on something that teachers use every day. days, often hundreds of times a day: intuition.

Sometimes called intuition, or sixth sense, even a Spider sense, intuition is the ability to read a situation and know something, without proof or conscious reasoning. It is "subtle knowing," writes Sophy Burnham in The Art of Intuition, "without ever having any idea why you know it."

In many professions, especially those that require lightning-fast decision making, such as firefighting or medicine, the ability to tap into our intuitive senses is critical. In 2015, the US Navy even began a program to investigate how service members could improve their intuitive skills during combat, following discussions with soldiers returning from deployment who said their intuitions often alerted them to imminent danger, even when information reliable data would not be available.

The teaching profession is no different. While teachers generally don't deal with dangerous or life-threatening situations, classrooms are complex and situations change quickly. As Anjali Nirmalan, Ed.M.'17, points out, "Intuition is extremely important for a teacher in a room full of other humans – in my case, over 30! - with a spectrum of your own needs. "

(Credits: Ed. Harvard Ed. Magazine – www.gse.harvard.edu)

Case 04 - Jeff Bezos and the role of intuition in decision making

By Sean P. Murray October 9, 2018

When it comes to decision-making, Jeff Bezos is comfortable trusting his intuition. That might come as a surprise, given Amazon's reputation for data analytics. Bezos has said in the past, "Our success at Amazon is a function of how many experiences we do per year, per month, per week, per day."

Judging by this quote alone, one might imagine that Amazon employees are like scientists in a laboratory, carefully tracking the results of experiments and analyzing the data to make every decision. However, this analogy would be misleading. While the culture of experimentation at Amazon is strong, there are some decisions that just don't lend themselves to data analysis. This is what Bezos said:

"All of my best decisions in business and life have been made with my heart and intuition - not through analysis. When you can make a decision with analysis, you should, but it happens in life that your most important decisions are always made with instinct, intuition, taste, heart."

(Credits to: www.RealTimePerformance.com)

Case 05 - Peanut.

In 2019, I went to buy raw peanuts at a small market.

As I had run out of packages on the shelf, I asked the market employee to weigh three packages of 500 grams each.

The first package gave the first weighing 500 grams just right. In the next one, I commented that it would give 503 grams, as a result, it gave 503g, in the third package, I said that it would give 518 grams, to the astonishment of the employee, a customer who was accompanying him and for me too, it gave 518 grams. The customer next door even asked me for the lottery numbers.

It may seem like double luck, but it was the values that came to my mind instantly, i.e., future information. Or to be more specific, action of Intuition.

Case 06 – Shirlei's friend.

On an afternoon in 2020, I was cutting the grass in the back of my backyard, while cutting, came the strong memory of my friend Shirlei who lives in Rio de Janeiro. After seconds of thinking, my wife comes out on the porch and comments:

- I think someone from your friend Shirlei's family died! I just saw "Luto" on her Facebook profile.

I thought it was her husband who had health problems, but talking to her, I learned that it was a friend of hers who lived abroad.

This type of phenomenon happens to many people, something even natural. Thinking of someone and soon comes the news of this person.

The question is how this phenomenon occurs and what is its importance in the scientific context.

We are talking about the psychic refinement of civilization, or perception through capturing the most varied situations in the form of waves. They are fragments of something much bigger that is to come for the benefit of human beings, and of progress in all areas of knowledge.

In this commonplace example, a deep and subtle science is present, the training of psychic mental potential for new climbs. This type of manifestation can occur for more important situations, such as in the scientific area, for example.

Sergio Antonio Meneghetti

"The pyramid of knowledge begins with the base stones".

Personal Preparation Tips

You may have noticed some of my behaviors leading up to intuition.

I am not going to dictate rules of how to obtain the best conditions of balance and harmony to understand intuition, but I can give you some tips that might help.

Every human being is in an evolutionary stage and also has his or hers particular affinities. Some feel calm with classical music, others with rock or samba, it doesn't matter, what matters is to feel good with yourself and to be attentive.

It is very important to begin to identify your thoughts in order to know what comes from your own mind and what comes from outside origin.

It is important to think logically, for instance: When we tune in a radio station and the transmissions is bad, of book we also get bad content. For this reason it is good to be tuned in a good station, to receive a good transmissions and also good content.

For you to ponder: did you ever notice that when we are anger with a person, we often have a real mental struggle.

Is it just our thinking?

Could it be that our mental radio station is picking up waves from other sources not so good?

On the other hand, have you noticed that when you are balanced, or performs goods things for your fellow beings your mind vibrates stronger and nice thought follows.

So is the human being. A real machine that interacts with all and suffers the most varied effects. The option is ours to choose what suits us best.

Observe that there is a logical science, in the abstract concept of thoughts, feelings and morals.

Did you stop to think that in fact, I am not just talking about demonstrating a psychic tool but also a merging science, religion and philosophy. And is this way showing with logic and fact, arguments so that "you" become a live tool, better and more refined. This is the secret to a good inspired intuitive reception.

I know it is not easy as I feel it on my skin and only with much effort is possible to go up the hill of our inner growth.

Realize that the job throughout it range of possibilities starts to take a more selfless characteristic and it is not simple something to earn your living.

What a wonderful way that gives us professional growth, personal development and so benefits that we love so much.

Here I am anticipating a trend that will be mandatory in

the future of all involved, the professional, the business owner, the entrepreneur, the student or the researcher.

With such a vision, you will be taking larger strides.

Choose a place for yourself, it can be at work, at home or any other place, making it a special place where you can elevate your spirituality.

No my particular case, I chose a place at work for this purpose and also a place at home.

Pray or simply think of good things in this place, it doesn't matter what creed it is, but the good intention.

This environment will suffer a positive action and you will be permeating the place with positive energies that will facilitate positive things. It is pure physics.

It is worth repeating something essential: <u>Reason disturbs intuition,</u> thus release your mind, let it go free to its natural ways and it will find the answers to your needs. It is normal for people to comment that they thought a lot about a problem and could not find the answer, but later when they were in a relaxed atmosphere or a leisure, the answer came to mind without these people even thinking about the subject.

Meditation is something that helps a lot to silence the mind and elevates the connection with goods things. In my case I don´t use this practice, but each person has his or hers characteristics and peculiarities.

About rules, I say that the only to be followed are those about your personal safety (physical integrity) and of others, and respect to all and everyone.   Otherwise you have freedom to do as you wish.

Some of the symptoms that happen to me and that it

might happen to you:

When I am going to do something and feel the desire to do it differently from the usual I let this feeling guide me. Often, with that new option I come across to situations and people that were important to me.

Something else, I observe my heart, when it is tight, it is because something not good is happening or could happen. Example: A certain time I was in a place and my heart was really uneasy, giving me a feeling of fear even. Result, I was mugged.

It may sound strange, but our heart is a compass the guide us. It is linked to the energy field where everything is connected, and where time and space lose their characteristics.

All of these that I comment are physical and chemical events and only require more attention and studies.

If you are an entrepreneur, think about your projects and release your mind. The best answers will come. You know better than I how much a good idea generated great businesses in recent years. Take the opportunity to get the best out of a situation. I am not talking about advantages, especially if you bring losses to others, but bring up ideas. In other words, note how the progress it brought reduced time and cost and business processes.

In the universe there is a law of economy generating preservation: to do more using the minimum of resources. Note that nature generally used only what is necessary and in the best possible way.

Staying calm is the best exercise to give vent to good ideas. I repeat: reason disrupts vision of new ideas.

In my work, I try to be as authentic as possible, that is, I create my own raw material to build something. As an example in the video I made to introduce his work, I used my photos of Sun rise and Sun set. I did the work of editing and montages in Photoshop. I am not a professional by I am exercising my mind all the time.

My greatest tool is the thinking and that has to be sharpen.

Generally the idea of a new business comes fragmented, as I commented earlier, the important thing is to understand the first fragment and work on the idea.

It is worth to remember that a plant comes from a tiny seed.

I want to one day receive the grateful news that my students were successful in their endeavors.

As of now, I ask God to help you in your life and in your efforts. Think that way.

If you are a **professional**, think of the welfare of the company and let the ideas flow. Don't let the ideas go to waste, have faith and courage, and present the ideas to the company. The thought without action does not generate progress. A robust and growing company ensures the daily bread to tis employees.

When I was an employee, I always was looking for practical ways to execute the same task, this regardless of anyone asking, it was for the simple pleasure of self-growth. The size of your company is not important, All the areas may experience improvements, if someone performs the same job better, adopt the same method as an addition to your learning that it will help you in the future with a better vision.

Thank God I was able to work in large companies: Lyondellbasell, Polibrasil, Chevron, Scania, (PE (Instituto de Pesquisas Energeticas e Nucleares), Oxiteno Quimica. So, I grew a lot within the structures of these companies.

Although not measured or proven, but all the energy that you put into your work producing something good, you can be "sure" , you will be generating a merit in the universe, and at some point this merit returns to you, whether in the same company or another occasion. It is law of physics, every action generates a reaction (physical causes generate moral effects, and moral causes generate physical effects). I learned that in "Great Messages" of Pietro Ubaldi and found it in practice in my existence.

If you are in the **development sector, keep in mind** that the major resolutions, new methods and products, should happen outside traditional route. Just so happens the new, otherwise it is just continuity or saturation of a process.

I would like to be at your side in moments of doubt, but as it is not possible, remember: **you too can be successful.** We are not aware of our full potential; our mind has the power to materialize situations. Using the gospel (without the mystical side of it), when Jesus spoke:

If you have faith the size of a mustard seed, and say to a mountain to move, the mountain will move.

In that sentence, that is a symbolism, Jesus is telling us of the potential of the mind and its power to manipulate matter, energy and situations.

Note that within pure religiosity, there is a deeper wisdom and science.

And all of this is in latency within the human being.

Do you agree that if you make a process or project quickly and cost-effectively, you will be saving multiple resources and so, generating progress in a sustainable way? This important action of yours will be contributing for a better world. Isn't this what life is all about?

When I am involved in a new procedure, I try to understand as much as possible the subject, I imagine myself going thru the whole process as if I was inside it, navigating throw each phase. The best way to know a phenomenon is looking and participating closely in the process.

To clarify, here´s an example:

As I worked with mixture of polishers in order to generate compounds with better properties, I imagined each resin melting inside the extruder. I imagined if mixing resins as if you were watching the event within the process. For those who don't know this kind of process, imagine yourself mixing butter with vegetable oil, it is hard if cold, but if you heat controlled, an ideal and homogeneous mix would occur, so your e products will have better features. Each product has its ideal characteristics, but not always the ideal it is what it is need for the how to do at hand. When I achieved my goals, those actions came from outside empiricism or progression of the traditional method. "Remember a saturation process technology", something X CD vinyl record, after CD X MP# and so on.

Joint the group and accept all ideas, even the most seemingly absurd. If come to a kind of an absurd idea and there resistance, but you feel strong about it, ,go ahead and try it, even putting yourself in situations of personal risk in relation to other members of the group or boss. **Not risk regarding safety or important rules not risk company.** Much of my achievements were obtained in stubbornness and anonymity, done at my own risk and after the good results I announced to those involved.

I performed several studies on gaps of time and anonymity, because, of book, I would have the "not" from the people responsible (hierarchy).

I should warn you: the world is ready to say "no" to something that you don´t understand, so, the pure scientist´s life science suffer. Going back to Einstein: he sent several letters about his theory, Max Planck only gave him attention, and it was still a struggle to get people willing to test his theory that was confirmed only years later.

When I speak of my person, it is not to be lauded, but I have to talk about the depth of my experience, in the sense that you may identify yourself with it and know that it is possible. When I relate something to you, it is because I have lived and tried the workings of intuition or psychic process.

If are **the owner of a company or business**, naturally you will need strategic vision, and quick decision making. Observe your heart imagining the possibilities ahead. Example:

You have two options, and only one is ideal for the future. Imagine yourself opting for condition "A" and note how you feel it in your heart. Then imagine opting for condition "B" and observe again how you feel it in your heart. The option transmits peace and tranquility in your heart will be the best then. When an option oppress your heart bring an air of fear and insecurity, get out, even if it seems a good thing at the time. It is important for a good analysis and common sense.

Remember that your emotional state is very important at this time. If you make decisions with a hot head, you will open to treacherous inspirations. Equilibrium is very important at this time.

You are the most powerful person in your business, and then you have the most complete view on it, so you have the autonomy to make decisions that can alter the direction of the company. Your intuition will carry out major changes in the whole. Remembering that your company does not have only the goal of generating profits, but progress and wellbeing of tens of thousands of people. You have a greater commitment with Creation, believe it or not. And all the good you do for the planet will have its reward, as well all the harm will have its effects. Remember, the first benefits from your success, are those who are the most expensive to your heat. We all want to do what´s best for those who are on our side.

Give freedom to all levels of the corporation, to produce and present ideas, without discarding the apparently absurd.

Don't listen to just your direct subordinates, but from the smallest to the largest in the hierarchical scale. Not always the good ideas arrive by usual channels, due to the nature of human beings. There are a lot of obstacles from one end of the hierarchy to the other. I repeat: go to the other end and come asking everyone what ideas they who have to improve the processes and the business.

Your first conquest will be the feeling of appreciation for the whole team. The result is the commitment of the majority to help. The second fact, you´ll have a satisfaction in your heart, an experience that you´ll never forget, financially as well as spiritually. It is practically impossible to separate work from feelings. Professionals react according to their feelings. It is pure logic.

If you are **a student**, you'll be increasing your luggage of how to act in the future. You´ll be anticipating a trend.

A suggestion to complement that study is my book: "**Intuition, Work Tool**" as in this book I give you tips on to improve the atmosphere in the work place, important for those who are still active.

In this work I have enhanced learning the inside out process, contrary to the process from the outside in. As soon a human being become aware of this fact, the soon he´ll be able to move forward.

Can you imagine the difference of you aware of this fact and those who still do not know him? I do not want to propose a dispute, but to alert you that you will be much better prepared.

You may have heard, or seen someone comment, what one learns in schools is only the results of the past. They are right. It is logical. You will always receive something chewed up unless it is invested to develop new projects. It will be in this new project that you will be able to make a difference with a new mindset and boldness.

As I said with other words, smart is not only get good grades and be able to answer most of the questions, but "also" have average grades and have ease resolving new issues and make them happen.

Life is made up of different thing so that the world may have all the pieces in their right places, that way it will work in harmony. No matter the function of each individual, everyone is important in the overall context.

In this work I am referring to various situations, and then try to extract the maximum amount of information for your growth while you are not active. Think of it as a professional, as the entrepreneur, and as a researcher, but above all, think of yourself, and keep in mind that you can accomplish great works in the future.

If you are a **researcher**, intuition is vital, because major discoveries came through the intuitive process, even with minds not having notion of this fact. If you accept what science proposes as rules (except immutable laws) you will be making little progress. Boldness is critical. Science walks through an alley and no matter how refined the instruments, and how profound it touches the science of pure phenomenon, there will always be questions. To study physical and chemical phenomenon in the field of energy and matter is one thing, but when science stumbles upon conceptual phenomenon, only a psychic and intuitive understanding can give answers. I go further into this subject. Intuition is, and it will be, the biggest search tool that human beings have in hands.

Examples speak louder, so again I am going to use the figure of the great genius to show you an important fact: If Einstein had not the boldness to face Newton on his theory of gravity or gravitation; the world would not have significant progress today.

See that there is much value in bold as the knowledge itself.

A note about official science:

Scientists in their great majority, dismiss religiosity for considering that it is full of "dogmas", so, it is so because that's how it is. Science itself contains more expressive dogmas because everything is accepted if tested in laboratory (be it inside, outside or even in space), but there are some many phenomenon that science can´t explain and do not cease to exist because of that. Another fact is that many scientists only know superficially the other side of the coin, and use their knowledge as parameter to approve or not the phenomenon, in other words, it becomes "true" if the scientist understands it and accepts it. Right now pride and vanity speak louder than science itself. You will hear many absurdities exposed by learned due their vision of the universe or other areas, but poor in other sources of knowledge. So, that imposes a question:

The great genius is he who knows a lot of experimentation and accumulation of information with his academic titles, or that scientist with little projection or knowledge discovers laws, other phenomenon factors that change the book of science and progress.

My intention is not to criticize this or that, but to show you to NEVER feel forced to follow a book apparently logical by science, but that you may have courage to follow by your own legs and ideas.

Another question for you to think about:

Current science took as true the evolutionary process that the universe took to get where it is. (I am not going into the merits of Evolution and Creationism).

If everything is the result of evolution, this means that everything evolves, correct?

Everything evolves, and then everything suffers this process, right?

Everything evolves, so the energy (no matter what) also has to undergo evolution, correct?

If this is true by the official science, could it be that energies generated billions of years as gravitation, or law of attraction and repulsion suffered evolution?

How about love and hate? Aren´t they energies with the same characteristics of attraction and repulsion?

There is another concept working and causing chemical reactions. You can only register the reflection of this concept and not the concept in his essence.

How to research or understand something that instruments do not have action on?

Only through the psychic perception can human beings understand the conceptual phenomenology.

How many times you felt a situation and didn't find words to express or explain the fact?

Don't miss the essence of pure research, launch hand <u>to all possibilities.</u>

Remember, I am not the owner of the truth, so research!!!

Reinforcing all that our senses and instruments registered are vibrating manifestations of the dynamic energy. This manifestation tend to the infinite, as certainly <u>there is a sea of manifestations that we don´t know of due the inability of our senses and instruments</u>. Never disregard the possibilities.

We only know a part of the building of the universe in the "relative", where manifestations occur in time and space. Imagine what lies ahead.

For the researcher, the only gate that must exist is the ethics.

With these concepts of freedom, I leave here a remark. Place the intuition and the synthesis ahead, ad let reason and analysis work on the information obtained.

Release your mind and let the free thought to uncover the phenomenon towards the infinite.

The expiation of consciousness will open a lot of doors:

One of them will be telepathy.

Thanks

Love Soul

Soul of my soul

Love my soul

That my loves you.

Child of light

Exults the divine work

Work! In this Divine work.

Launch yourself into the universe

And the universe will receive you

Grow in this universe

This one, called Love.

Sergio Antonio Meneghetti

Sweet caresses will come to you

And the light of the soul illuminates

The light of other souls.

Come on wisdom that saves

Save your brother in wisdom

Universal wisdom that love

Love my soul

Soul of mine

Love with your spirit

All souls.

(Sergio Antonio Meneghetti)

I thank all the souls who fought and are fighting for the light of knowledge.

About the author

Sergio Antonio Meneghetti

São Paulo – Brazil

Intuitive Scientist, Writer, Speaker and Chemist.

Universal Ambassador of Peace – France – Geneva – Switzerland – Cercle Universel des Ambassadeurs de la Paix.

• Member of the International Association of Poets

• Member of the Peace Movement – Poetas Del Mundo

• Member of the Fondation Franz Liszt – France

Author of the books:

• Intuição Ferramenta de Trabalho – Self-development

• Intuition Working Tool - Self-development – English version – USA

• Intuição para Mulheres – Self-development

• Multiplicando a Genialidade

• Multiplying the Genius Within - English version

Sergio Antonio Meneghetti

- Multiplicando la Genialidad – Spanish version

- A Intuição no Avanço da Ciência e Tecnologia

- Intuition in the Advancement of Science and Technology – Self-development

- A Reconstrução do Universo - Science.

- The Reconstruction of The Universe – English version.

- Do Hidrogênio ao Hélio – Sem Fusão Nuclear – Science

- The Quantum World and the Expansion of the Universe - Cosmological Model by Vortices - participation Dr. Lior Shamir from Kansas State University – Science

- O Fim SWem Fim do Universo – Science. • Freedom of Conscience - Philosophy

- A Construção do Pensamento - Philosophy

- Homem de Barro – Philosophy

- O Sertanejo de Goiás - Fiction Novel

- Vida de Água - Fiction Novel

- Gestão é Uma Arte - Human Management

- For Those Who Work in New York - English version - Career

- Socialmente Falando - Sociology

- Paz no Mundo – Volume I – Poems

- Paz no Mundo – Volume II – Poems

- Sem Saber Sabino - Stories

- O Cavalinho Dourado - Children's

- O Pequeno Florista - Children's

- Emilião - Children's

Author of Scientific Hypotheses through Intuitive Psychic Perception:

• Birth of Hydrogen and growth to Helium.

• Formation of Subatomic Particle

• Curvature of Light at great distances (Curved Expansion of the Universe).

Interviews:

• Vanguarda TV

• Rede RVC TV (3)

• Band Vale TV

• AllTV- SP

• TV Taubaté

• Think TV

• Tatiana Fedatto (2)

• Agoravale

• Acontece Pinda

• Corre Certo Program

• Rádio Difusora

• Rádio Rede Assim (3)

• Rádio Princesa

• Rádio Vale FM

• TVI – S. J. Campos

• Proza.Podcast

Lectures:

• AJOP – Pindamonhangaba Journalists Association.

• Therapeutic and Artistic Space (How to Overcome Industry 4.0 and Artificial Intelligence).

• IBIS Taubaté Hotel (Intuition in Your Profession)

• Nova Gokula Pindamonhangaba (Intuition, the psychic tool of the future)

• Dr. João Romeiro Pindamonhangaba College (Intuition in Business)

• Anhanguera Taubaté College (Intuition in Business)

• Pindamonhangaba College FAPI (Intuition in Business)

• Anhanguera Pindamonhangaba College (Intuition in Business)

• Spiritist Center: Love and Charity – Orlando – Florida – United States.

"Nexis for Leaders (NPL)" course by UPPER Aducation.

Teachers: Walter Longo – Flávio Tavares – Zé Luiz Tavares – Thiago Nigro – Jaime de Paula – Ricardo Nunes Kiko Kinslansky – Marcelo Molnar – Robson Henrique – Sergio Antonio Meneghetti

Jobs:

• Lyondellbasell (Formerly Polibrasil)

• Chevron Química do Brasil

• Institute of Energy and Nuclear Research – IPEN

• EMCA

• Atlas Indústrias Químicas (Oxiteno)

• SAAB SCANIA

Acknowledgements Received:

• Barack Obama - Presidential Email (04/02/2010).

• Pope Francis's Acknowledgment to the Cercle Universel des Ambassadeurs de la Paix

• Honorary Merit for Humanitarian Work in Favor of Culture and Peace – Zap Magazine - 2009.

• Robson Miguel – World's No. 1 Guitarist

• 2013 Poetic Highlight Award – ALAF (Fortaleza Academy of Letters and Arts)

• Ayrton Senna Institute (in the name of Viviane Senna)

• Jardim Pueri Domus Unit

• Rádio Nova Brasil FM

• Doutores da Alegria

• David Feffer "Suzano Group".

• Volker Trautz (CEO) international "LyondellBasell Industries"

• Featured of the Month in Polibrasil

• Pão de Açúcar Cable Car.

Participations:

• Magazine: Segredos da Mente – articles about intuition in 3 editions.

• Invited to participate: UN – BRAZILIAN PEACE, LITERATURE, SUSTAINABILITY AND ARTS – 2016.

• Brazilusa Magazine Orlando (USA) – columnist.

• Tribuna do Norte Newspaper

• Agora Vale – Column – Work Intuition Etc. – Pindamonhangaba

• Participation with articles and poems on the following websites and

Sergio Antonio Meneghetti

newspapers:

- www.administradores.com.br/sergio59

- Dia-Dia-News

- Pensador – UOL website

- Vale Empresarial

- Rádio Raizonline – Portugal

- Exemplar Magazine – columnist – Pindamonhangaba

- Contemporary Literary Horizon – Romania

- ABC Chemists' Union Magazine

- Rádio Mundial

- Villagenews Newspaper – Pindamonhangaba

- Condomínio News

- STOP the Destruction of the World (International NGO founded in Paris – France) www.stop.org.br

- SITA – International Society of Analytical Trilogy

- Café Cultural – SESI – Santo André

- Jornal da Cidade – Pindamonhangaba

- Jornal do Brasil – Rio de Janeiro.

- JB Online – Rio de Janeiro. • Jury member at Festipoema 2010 – 2022 - 2023

- Participation in the Exhibition – Black Consciousness – Pindamonhangaba Museum

Honors Received:

- Honoree in 2017: Hellen Morais Raybbot Gonçalves – graduate in Administration.

- Motion of Congratulations from the City Council of Pindamonhangaba -

2008.

Accolades and participation in poetry competitions (books):

• Introduction: Cape Verde – The Other Side of Politics (Carlos Fortes Lopes)

• Preface: Versos Soltos (Carlos Fortes Lopes – Cape Verde)

• Anthology of Brazilian Poets volume 5.

• II Cultural Olympiad – "500 Years of the Portuguese Language" 2005

• III Cultural Olympiad – "500 Years of the Portuguese Language" 2006

• Golden Book of Brazilian Poetry

• IV Poetry, Short Stories and Chronicles Selection of Barra Bonita.

• Literary Panorama 2005/2006 (6500 entrants).

• New Poets New Talents

• Poets of Brazil

• International Competition of the Voz Di Studanti website - (Cape Verde).

• 4th Literary Contest of Short Stories and Poetry

• Poets of the World in Poetry – volume I

• Anthology of the Pindamonhangaba Academy of Letters (2012)

• Anthology "Mulheres Entrelaçadas" (Launched in Germany)

• Electronic anthologies: Fenix (Portugal) and Editora Pragmatha.

• Anthology Messages for the Future – 2020.

• Anthology The Return – 2021.

• Anthology of the Pindamonhangaba Academy of Letters (2022)

• Anthology – Florilégio (2025)

E-mail: sergio.livro07@gmail.com

Sergio Antonio Meneghetti

www.ingramcontent.com/pod-product-compliance
Lightning Source LLC
Chambersburg PA
CBHW021415210526
45463CB00001B/382